FREE CREATIONS OF
THE HUMAN MIND

FREE CREATIONS OF THE HUMAN MIND

The Worlds of Albert Einstein

Diana Kormos Buchwald
and Michael D. Gordin

OXFORD
UNIVERSITY PRESS

OXFORD
UNIVERSITY PRESS

Oxford University Press is a department of the University of Oxford.
It furthers the University's objective of excellence in research, scholarship,
and education by publishing worldwide. Oxford is a registered trade mark of
Oxford University Press in the UK and certain other countries.

Published in the United States of America by Oxford University Press
198 Madison Avenue, New York, NY 10016, United States of America.

CIP data is on file at the Library of Congress

ISBN 978–0–19–767819–0

Printed by Sheridan Books, Inc., United States of America

In memory of
Ady Barkan
1983–2023

CONTENTS

Acknowledgments | ix

Prologue: 1921 | 1

1. Bern and Princeton | 9

2. Relativity Theory | 29

3. Quantum Theory | 49

4. Belonging | 69

5. War and Peace | 89

6. Free Creations | 106

NOTES | 115
FURTHER READING | 123
INDEX | 127

ACKNOWLEDGMENTS

We are grateful to all those who helped make the writing of this book possible. Our greatest debt is to the voluminous Einstein scholarship that has emerged over the past fifty years, bringing important documents and episodes to light. We owe sincere thanks to these researchers, who can only partially be acknowledged by name in the Further Reading section. In particular, the staff of the Einstein Papers Project at Caltech was very helpful in enabling the smooth functioning of a cross-country, and eventually trans-Atlantic, collaboration in the writing.

Constructive comments on a draft manuscript were provided by Jed Z. Buchwald, Margaret Leslie Davis, Erika Lorraine Milam, Patt Morrison, and John D. Norton. We greatly miss the wit and acumen of John L. Heilbron, who passed away in October 2023. One of his final missives to us was a characteristically astute reading of the text.

Finally, we thank the two anonymous readers for Oxford University Press and especially our editor, Nancy Toff, and our copyeditor, Betty Pessagno. They have all decisively improved the pages that follow.

Prologue: 1921

A man in a faded gray raincoat and a flopping black felt hat that nearly concealed the gray hair that straggled over his ears stood on the boat deck of the steamship Rotterdam *yesterday, timidly facing a battery of cameramen. In one hand he clutched a shiny briar pipe and with the other clung to a precious violin. He looked like an artist—a musician. He was.*

But underneath his shaggy locks was a scientific mind whose deductions have staggered the ablest intellects of Europe. One of his traveling companions described him as an "intuitive physicist" whose speculative imagination is so vast that it senses great natural laws long before the reasoning faculty grasps and defines them.

The man was Dr. Albert Einstein, propounder of the much-debated theory of relativity that has given the world a new conception of space, and time and the size of the universe.

Dr. Einstein comes to this country as one of a group of prominent Jews who are advocating the Zionist movement and hope to get financial aid and encouragement for the rebuilding of Palestine and the founding of a Jewish university. He is of medium height with strongly built shoulders, but an air of fragility and self-effacement. Under a high, broad forehead are large and luminous eyes, almost childlike in their simplicity and unworldliness.[1]

Thus, on its front page, did the *New York Times* begin the forging of Einstein's international persona, upon his first arrival to the United States in April 1921. Interviewed onboard ship during a brief required quarantine, Einstein was asked about opposition to his theory of relativity. He forcefully declared that "the attacks in Berlin" against the theory were entirely antisemitic. Even though Einstein had arrived with a delegation of Zionists to raise money for establishing a Hebrew University in Jerusalem in British Mandate Palestine, the journalists proved themselves well attuned to events in Germany. Einstein and his work had been the subject of venomous criticism from right-wing quarters back home, and this too—along with the promise of collecting significant honoraria from a handful of universities—was a reason for his visit.

Einstein was greeted in New York with avid interest from the press and enthusiastic acclaim from the Jewish community. He was surrounded by reporters, photographers, and "moving picture men" who filmed him aboard ship for more than half an hour. The journalists coaxed anecdotes about Einstein from his wife, Elsa Einstein, and from Chaim Weizmann, leader of the Zionist Organisation (and sponsor of this trip) and later the first president of Israel. Several hundred onlookers lined the Battersea pier, where the delegation was welcomed by Jewish leaders. There followed a reception by Mayor John F. Hylan in his offices, while the *New York Times* reported, "more than 5,000 Zionists filled the plaza" in front of City Hall.

Einstein's arrival with a Zionist delegation was a novel event for both parties, who met in person for the first time onboard. Einstein, a thoroughly secular Jew, had first publicly supported the cause of persecuted Eastern European Jews after World War I with a statement in the popular daily newspaper *Berliner Tageblatt* on December 30, 1919. In his remarks, he decried official intentions to take severe measures against East European Jews living in Berlin, who were increasingly being blamed for economic and political troubles, as a "horrible injustice."[2]

Tens of thousands of displaced Jews settled in the German capital after World War I, and their ubiquity made them easy targets for the brewing discontent of a defeated country feeling the economic squeeze of paying reparations dictated by the Versailles Peace Treaty. Over the next two years, his commitment to Jewish causes intensified, particularly in early 1920 after the tumult that arose during a physics class. Einstein had tried to get around the restrictions placed on foreign students, most of them Jews, from registering at the university, so he allowed some to audit his lectures on relativity; the Berlin University student council protested. Despite Einstein's efforts to minimize the political ramifications, the press described the uproar as antisemitic. The incident led Einstein to champion immigrant Jews in Germany, and eventually to his conclusion that a Hebrew University in Palestine, a place that would allow Jews unrestricted access to higher education, was a necessity. In the years ahead, this goal connected him to the conversations and advice of the Zionists Kurt Blumenfeld and Hugo Bergmann. It was through their offices that Weizmann invited Einstein to accompany him on a fundraising tour to the East Coast of America. Einstein agreed "without giving it more than five minutes of thought," he recalled.[3]

Einstein's renown was quite recent, descending upon him when he was forty years old. There would have been no uproar at a Berlin lecture with 1,500 attendees, and there would have been no invitation to the United States, had it not been for the announcement in late 1919 from a British astronomical expedition. Arthur S. Eddington, the leader of that group, trumpeted the confirmation of one of the central predictions of Einstein's general theory of relativity. It was true: the angle of deflection of starlight reaching telescopes on Earth as it passed close by a massive gravitational object, such as the Sun (as observed during an eclipse), was measured to be twice as large as the angle previously predicted by Newtonian gravitational theory (under the assumption of light as a mass-bearing particle). This result, together with Einstein's correct recalculation of the unusual trajectory of the planet Mercury around the Sun, constituted two of three significant observational findings that would distinguish the Einsteinian theory from its Newtonian predecessor. Suddenly, in late 1919, Einstein's portrait appeared in the German daily illustrated press, with headlines applauding him as a new "Copernicus, Galileo, or Kepler" whose work had upended Newton's universe.[4] Both *The Times* of London and the *New York Times* now featured the previously unknown Einstein.

The vindication of his work carried an internationalist aura. Eddington, a Quaker pacifist, had persuaded the British Royal Astronomical Society to sponsor an expedition that ultimately confirmed the theoretical work of Einstein, a German. Such observational verification had been attempted in the Crimea in 1914 by the German astronomer Erwin F. Freundlich, but his equipment was seized by Russia at the outbreak of war. International scientific correspondence ceased, conferences were canceled, scientific journals no longer circulated across the lines of combat. The official boycott of German scientists by the victorious Entente, and the sidelining of dominant German-language scientific journals and organizations, began only in 1919. This boycott was not rescinded until 1926, when—through no small efforts by Einstein—Germany entered the League of Nations. Reestablishing international scientific relations became Einstein's second most important political and professional activity after the end of the war, and the trip to the United States was his unique opportunity to reconnect with the Anglo-Saxon scholarly world that was still, in 1921, closed to German and Austrian scientists and intellectuals.

Einstein had other reasons to travel to America. As the journalists in New York harbor correctly diagnosed, Einstein had earlier considered the criticisms of his theory an aberrant abstraction. By 1918, in an article modeled on Galileo Galilei's *Dialogue Concerning the Two Chief World Systems* of 1632, he had joined the fray. The assault continued. The German journalist Paul Weyland, taking his cues from earlier attacks on Einstein's work by the experimentalists Ernst Gehrcke and Philipp Lenard (the latter the 1905 Nobel

Laureate in Physics), created an organization called the Working Association of German Natural Scientists for the Preservation of Pure Science. Weyland's main message was that Einstein's theory was nothing more than a "mass suggestion" that needed to be purged from science. At a special meeting of his organization in the Berlin Philharmonic Hall on August 24, 1920, Weyland accused Einstein—who was in the audience—of scientific dadaism, publicity-seeking, manipulating the press, and plagiarism. More troubling was the antisemitic literature distributed during the meeting and the swastika lapel-pins for sale at the entrance to the hall. The event received broad newspaper coverage, with headlines such as "The Battle Around Einstein's Relativity Theory" or "The Offensive Against Einstein." The distinguished physicists Max von Laue, Walther Nernst, and Heinrich Rubens wrote a short statement in support of Einstein, while headlines proclaimed that "Albert Einstein Wants to Leave Berlin!" Heartening private letters poured in to Einstein's home from such prominent colleagues as Arnold Sommerfeld, Hendrik A. Lorentz, Ernst Cassirer, and Fritz Haber. Several days later, the German minister of education published an open letter addressed to Einstein in which he decried the "malicious public attacks" that verged on libel. He stressed that Einstein had a "permanent place in the history of science," and he encouraged Einstein, who was "among the most brilliant jewels of science," not to leave the Prussian Academy and Berlin.[5]

In a daily newspaper, Einstein forcefully exposed the antisemitic undercurrent of the purportedly scientific critiques, much to his colleagues' consternation. He wrote to his close friend, physicist Paul Ehrenfest, that he "had to do this if I wanted to stay in Berlin, where every child knows me from photographs. If one is a democrat, then one must grant the public this much right as well."[6] He challenged bona fide antirelativist physicists to bring their arguments to the upcoming professional meeting of the Society of German Natural Scientists and Physicians, to be held in Bad Nauheim that fall. The leadership of the German Physical Society was divided in its support for Einstein, and friends and family feared that further publicity would harm both Einstein and his colleagues. (The meeting turned out to be less contentious than anticipated.)

By the end of 1920, Einstein had received several invitations to lecture in the United States. Eager to secure the financial future of his adolescent sons in Switzerland, Einstein (unsuccessfully) demanded exorbitant honoraria of $15,000 from Princeton and the University of Wisconsin, trusting that the land of "Dollaria," as he wrote, would be able to afford them.[7] His involvement with his colleagues in a progressive political group, the Bund Neues Vaterland, also thrust him into the news. Statements, signed by the Bund, that appeared in French newspapers regarding potential German violations of the disarmament prescribed by the Versailles Treaty led to accusations of treason against Einstein and his confederates, and to a call to assassination in a right-wing German newspaper.

Einstein felt a need to get away, even to a country that had not yet normalized relations with Germany. (President Warren G. Harding would sign the formal peace treaty only in October 1921.) These invitations to the United States, England, and France during the boycott alienated other German scientists, including his friend Fritz Haber, who rebuked Einstein for traveling at a time of great need for Germany and German science. There was by now no doubt that Einstein's internationalism, pacifism, and outspokenness set him apart, for better and for worse, from German academic circles. His answer: his trip to America would, on the contrary, lead to reconciliation among scientists and intellectuals. To a certain extent, Einstein achieved almost everything he set out to do during his trip: he provided a charismatic leadership figure to cultural Zionist endeavors; he forged personal and professional bonds with American scientists; he offered hope for peace and better times; and he presented his new theory to both academic and public audiences eager to learn about the new physics.

Perhaps the most significant result of his trip was the series of lectures he gave at Princeton. They would be published in English

Einstein converses with Henry Burchard Fine, Princeton University's dean of the Departments of Science, at City College, New York, during the physicist's first visit to the United States in April 1921, amid a crowd of onlookers. Princeton University would later name the Physics and Mathematics building in Fine's honor. Einstein would have an office there in the 1930s until the facilities at the Institute for Advanced Study were completed. *Courtesy Seeley G. Mudd Manuscript Library, Princeton University Library.*

shortly thereafter as *The Meaning of Relativity*, a book that remains in print as one of the most important expositions of the subject. During his stay in the United States, Einstein was repeatedly asked to comment on ongoing experiments intended to disprove relativity, particularly those that challenged his postulate of the constancy of the speed of light and the nonexistence of a luminiferous ether. While still at Princeton, Einstein had learned about preliminary experiments undertaken at Mount Wilson by Dayton C. Miller of the Case School of Applied Science, who claimed to have found a non-zero ether drift, in contradiction to Einstein's theory. As over-heard by Princeton's mathematics professor Oswald Veblen, Einstein is said to have responded: "Subtle is the Lord, but malicious

He is not." A decade later, Veblen asked Einstein for permission to inscribe this "child of your wit" over the fireplace of a new Department of Mathematics building. Einstein agreed. It was in this building that Einstein, now in exile, would have his first office in America only a few years later.[8]

Chapter 1

Bern and Princeton

Albert Einstein preferred to be rooted in place. His life began with a series of peripatetic disruptions: he was born in the southern German town of Ulm on March 14, 1879, but was moved to Munich the following year, where he attended most of his schooling, followed by sojourns in Pavia and Milan, Italy, and several cities in Switzerland. Yet from the age of thirty-five, in 1914, until his death in 1955, he based his life in two places: Berlin, Germany, and Princeton, New Jersey.

Depending on where you find him, who Einstein was could be remarkably different. Einstein in Berlin—a topic on which several books have been written—lived at the center of public and scientific life. His colleagues pulled out all the stops to induce him to move in 1914 to the German capital from his alma mater, the Eidgenössische Technische Hochschule (ETH; Federal Institute of Technology) in Zurich, Switzerland. He accepted a triple position in the bustling metropolis: member of the Prussian Academy of Sciences, professor at the University of Berlin (with the right to teach as much or as little as he desired), and director of a putative new Kaiser Wilhelm Institute for Theoretical Physics. Even before he was catapulted to international fame in the wake of World War I, he was much sought after as a leading light of the capital. This adulation only accelerated in the 1920s.

The uniqueness of his Berlin circumstances can be appreciated by contrast with his first full professorship in theoretical physics, at the German Charles-Ferdinand University in Prague, which he occupied from April 1911 to August 1912. As a new arrival, his social circle was fairly circumscribed in the third largest city of Austria-Hungary, and he dedicated himself almost exclusively to seeking the proper principles for his theory of gravitation. Although his social network overlapped with that of a young writer named Franz Kafka, and the two likely met on one occasion, lovers of historical coincidences will be disappointed to learn that neither of them ever recalled the encounter.

Although interesting engagements and near-misses can be uncovered for every place where Einstein spent significant time, two locales stand out for their outsized impact on both his life and his public image, one at the beginning of his career and the other at the end.

The first locale is the small city of Bern, the capital of Switzerland, where Einstein lived from February 1902 until October 1909. This Einstein resembled the lower-level bureaucrat he in fact was: Patent Officer, Second Class, clad in an affordable suit with well-trimmed dark hair and a neatly clipped mustache. During one remarkable year, this unassuming twentysomething with a young family altered the direction of physics forever. This "young Einstein" has been immortalized as the quintessential scientist. At the other end of his life, in the even smaller town of Princeton, we encounter a quite different visage. The mustache had grown bushy and, together with the shock of hair—sometimes unkempt, sometimes raked back—had turned gray and finally white. The clothing was idiosyncratic, characterized by a sweatshirt, a sweater, or a leather jacket, and the scientist was sometimes in sandals and sometimes sockless. This is the Einstein that now occupies countless posters and memes, captioned with philosophical (and often incorrectly attributed) bon mots.

The Einstein of Princeton did not flinch from arguing strongly in favor of pacifism and Zionism, against antisemitism and Nazism, and was consumed by the specter of nuclear weapons. Yet physicists today no longer associate this Einstein with much durable physics. The science that Einstein considered his most important in Princeton, from his challenges to the dominant interpretation of quantum mechanics to unified field theory, has been dismissed as the misguided effort of a genius past his prime. "Old Einstein" appears more like a sage than a scientist.

Like many popular representations of Einstein, there is much that is exaggerated in both the "young Einstein" of Bern and the "old Einstein" of Princeton. Most striking in retrospect are the continuities that illuminate consistent themes of Einstein's career. It is true that his explosion of scientific creativity in Bern was unmatched in impact compared to the works Einstein published in his later years. Indeed, they are unmatched in the work of any modern physicist in the century before or the century since. It is also true that as a Nobel laureate and the world's most visible refugee in the 1930s, his public role in the United States was much more prominent than that of the Swiss technical clerk. Nonetheless, intellectual concerns, struggles with his employers, and family life all link together the young and the old Einstein.

The Young Civil Servant

At the age of twenty, Einstein became Swiss by choice, filing his application to become a citizen of the canton of Zurich in 1899; he had rescinded his German citizenship—or, more accurately, as a minor had his father do so on his behalf—three years earlier. The young man moved to Switzerland from the environs of Milan, where his parents and younger sister had just relocated upon the failure of the family's electrical engineering firm in Munich. His heart was set on studying physics at the Zurich Polytechnic (ETH). To this end, he spent a year preparing for his entrance exams in the town of Aarau, within 50 kilometers by train from his goal. He enrolled the following year and graduated in 1900, having made a friend for life in Michele Besso and after falling in love with Mileva Marić, both fellow physics students. His Swiss citizenship was granted in February 1901.

Einstein's career prospects were nebulous. Upon graduation he was unable to secure an *Assistent* position, the first rung on the academic ladder. He resolved to stay in Zurich, near Marić, earning money by giving private lessons while beginning a doctoral dissertation. After another failed round of applications in 1901, the father of his classmate and future collaborator Marcel Grossmann interceded with the head of the Swiss Patent Office to give Einstein a paying job. Einstein held down a temporary teaching position at the Winterthur technical school until the Patent Office appointed him as an examiner in mid-June 1902. He started work a week later.

It proved an excellent fit. He had long been fascinated by electrical technologies like those his father and uncle Jakob had made the core of the family electrical manufacturing business. He discovered he had a flair for evaluating the originality of the inventions flooding the office during this era of electrical ingenuity. His interest was not merely a temporary adaptation to circumstances; in later years Einstein worked as an expert witness in patent disputes and filed quite a few patents of his own even after his reputation as one of the world's leading physicists was assured. His superiors in Bern noted his capacities: he was elevated to a permanent position in September 1904. That same year he was joined by his friend Besso, who was hired upon his recommendation.

Outside office hours, his scientific and social worlds also continued to expand. In 1902, with his student friends Conrad Habicht and Maurice Solovine, he formed what they dubbed—with self-mocking, ironic grandiosity—the "Olympia Academy." This discussion circle, complete with an overly elaborate diploma of membership, was the vibrant core of his social life. Besso joined in due course. The group read and discussed artistic literature (such as *Don Quixote*) but principally concentrated on the philosophy of science. Prominent on the reading list were figures such as Karl Pearson, John Stuart Mill, David Hume, Baruch Spinoza, Hermann von Helmholtz, Bernhard Riemann, Richard Dedekind, and Henri Poincaré. Most salient in Einstein's own recollections were the books on mechanics and thermodynamics by the Austrian philosopher and physicist Ernst Mach, which argued for the elimination of any concepts in physics that could not be anchored in rigorous empirical foundations. Aspects of Mach's style of reasoning characterized the astounding breakthroughs Einstein accomplished during his Bern years.

His closest companion was Marić, whom he married on January 6, 1903, over the strident objections of his parents, who disliked the older, moody, non-Jewish, and partially lame young lady as a soulmate for their son. In January 1902, a year before the wedding, Marić had given birth to a daughter, referred to in the couple's correspondence as "Lieserl," who was left with Marić's relations in Novi Sad (Serbia), and then vanished from the historical record. She likely died of scarlet fever in childhood. On May 14, 1904, their first son, Hans Albert, was born; a second son, Eduard, was born in Zurich on July 28, 1910. In Bern, the couple changed apartments every few months, adjusting to the limitations of Einstein's salary.

Tensions in the marriage worsened as the years progressed. Marić had been admitted to the ETH in the physics section at a time when female students were a rarity even in the comparatively tolerant Swiss universities. (The ETH officially permitted women to study as early as 1855, but the first to matriculate, a Russian named Nadezhda Smetskaia, enrolled in mechanical engineering only in 1871.) Before their marriage, as Marić struggled in her studies, Einstein's letters to her were encouraging. These have occasionally been interpreted as evidence that she closely collaborated on his seminal Bern papers of 1905, although neither she nor any contemporary made such a claim for credit during her lifetime. The strains in their marriage seem to have been of the ordinary kind: the dimming of

Einstein's Zurich family in 1929 stands on the balcony of their apartment at Huttenstrasse 62: his sons Hans Albert and Eduard, his first wife Mileva Einstein Marić, and Hans Albert's wife Frieda Einstein-Knecht. *Photo by Ljubomir Trbuhović, courtesy Stadtarchiv Zürich.*

affections, managing a growing family, and experiencing the nov-elty of Einstein's rising professional recognition.

That recognition came in large part from the extraordinary flourishing of creativity that Einstein displayed in a series of papers penned during evenings and weekends across the year 1905. His-torians call it his *annus mirabilis*: the wonderful year. By this point, Einstein had already published five papers. None of these publica-tions would have alerted readers to what was to come following his twenty-sixth birthday. In nine months, he wrote an astonishing series of six papers, including his doctoral dissertation, that would engage physicists, astronomers, and other scientists for the next century. This phenomenal productivity does not even include the twenty-one diverse reviews he published that year on the current physics literature. The International Union of Pure and Applied Physics would designate 2005 the International Year of Physics in recognition of the labors of this Swiss patent clerk.

In May 1905, Einstein wrote a humorous letter to his friend Conrad Habicht, upbraiding him for a long silence with raw enthusiasm:

> So what are you up to, you frozen whale, you smoked, dried canned piece of soul or whatever else I would like to hurl at your head, filled as I am with 70% anger and 30% pity [...] after you so cravenly did not show up on Easter. But why have you not sent me your dissertation? Don't you know that I am one of the 1½ fellows who would read it with interest and pleasure, you wretched man?

And then he casually proceeded:

> I promise you four papers in return, the first of which I might send you soon, since I will soon get the complimentary reprints. The paper deals with radiation and the energy properties of light and is **very revolutionary**, as you will see if you send me your work *first*. The second paper is a determination of the true sizes of atoms from the diffusion and the viscosity of dilute solutions of neutral substances. The third proves that, on the assumption of the molecular theory of heat, bodies on the order of magnitude 1/1000 mm, suspended in liquids, must already perform an observable random motion that is produced by thermal motion; in fact, physiologists have observed <unexplained> motions of suspended small, inanimate, bodies, which motions they designate as "Brownian molecular motion." The fourth paper is only a rough draft at this point, and is an electrodynamics of moving bodies which employs a modification of the theory of space and time; the purely kinematic part of this paper will surely interest you.[1]

These bold claims were no exaggeration; they might even be characterized as comparatively modest, given that he considered only the first paper to be "revolutionary."

The Black Body and the Wonders of 1905

On March 17, 1905, Einstein submitted to the *Annalen der Physik*, the flagship journal of the German physics community, a scientific paper entitled "On a Heuristic Point of View Concerning the Production and Transformation of Light," which the journal's editor, Max Planck, published in early June. This was a fortuitous pairing, as the article addressed a series of papers published by Planck himself, then a forty-two-year-old professor of theoretical physics at the University of Berlin and member of the Prussian Academy of Sciences. Planck's recent work was a continuation of his early interest in the laws of thermodynamics, the science of heat and energy conservation central to the second industrial revolution. The significance of Planck's introduction of the "quantum" of action in 1900 was clarified only in the coming years, mostly thanks to Einstein's insights expressed in this radical paper of 1905.

The exploration of thermodynamics in solutions had been Planck's favorite topic for decades, but in the 1890s he hoped to reach a deeper understanding of a system's entropy—the term for the lost work of the dissipated energy, as coined by Rudolf Clausius in 1865, or for the system's disorder, according to Ludwig Boltzmann—so he turned to other physical systems: so-called black bodies. These are any cavities, such as ovens or stars, that absorb all incident electromagnetic radiation, including light; they also emit electromagnetic radiation in the form of heat and light. In thermal equilibrium with their surroundings, black bodies emit what they absorb. Yet Planck and others found that black bodies appeared not to conform strictly to the laws of thermodynamics as then understood. The light emitted by such systems depended strangely on the body's temperature: at high temperatures, the light emitted would find its peak in the visible light spectrum, but as temperatures dropped, the peak of radiation moved to the infrared spectrum, invisible to the human eye. Planck sought a theoretical interpretation "that had to be found at any cost, no matter how high."[2] He at first proposed a theoretical formula assuming the existence of tiny oscillators at the black body's surface, but his experimentalist colleagues at Germany's bureau of standards could not corroborate it. Undaunted, over the next two months Planck engaged in the "most strenuous work of my life," trying to conceive of a uniform theory of the behavior of black bodies, regardless of their temperature.[3]

Following on the work of his Viennese counterpart Ludwig Boltzmann, Planck developed a formula that gave a precise relationship between entropy and the level of molecular disorder by explaining how energy was distributed among the postulated oscillators. Planck introduced the notion of the total energy of a system as constituted by countable, finite parts. The energy of Planck's tiny oscillators was a product of this constant and the frequency (v) at which they vibrate, $E = hv$. You could have $7\ hv$ or $981 \times 10^{29}\ hv$; but you would never find a noninteger number like $8.3\ hv$.

Planck introduced a new natural constant, h, to mark out this discreteness as "the essential point of the theory," and led to a new radiation law that works empirically, as described in his *Lectures on the Theory of Heat Radiation* published six years later:

> An immediately striking feature of [the relationship $E = hv$] is the entry of a new universal constant h of which the dimensions are a product of energy and time. It marks an essential difference from the expression for the entropy of a gas [. . .] The thermodynamics of radiation will therefore not be brought to an entirely satisfactory conclusion until the full and universal significance of the constant h is understood. I should like to label it the "quantum of

action" or the "element of action" because it has the same dimensions as the quantity to which the Principle of Least Action owes its name.[4]

For the next four years, only a few physicists concerned themselves directly with the core issues in Planck's new distribution law, which at this point he himself understood to be purely a mathematical fiction, useful for calculation. Einstein had first mentioned Planck's radiation theory in a 1904 paper on the molecular theory of heat; the following year, in his light quantum paper, Einstein went further, making a powerful analogy with dilute solutions and ideal gases. Although voiced in a careful, moderate way as a heuristic point of view, Einstein's assumptions and conclusions about the existence of particles of light were extraordinary. In March 1906, he further showed that the radiation of each of Planck's oscillators was quantized, rather than the energy across frequencies being broken up merely for computation's sake—that is, that Planck's "quanta of action" were real, physical aspects of nature. Quantum theory was born.

Einstein's approach could explain for the first time puzzling empirical findings such as the photoelectric effect: the emission of electrons from a material object when struck by light or other electromagnetic radiation. (This is how solar panels work.) The effect was first observed and analyzed by Heinrich Hertz, the discoverer of radio waves, in 1887, and over the next years it spurred a flurry of laboratory work. Fifteen years later, the German physicist Philipp Lenard—who later as a vociferous Nazi would play an infamous role in Einstein's life—found that the energy of the emitted electrons changed depending on the color of the light hitting the surface of a metal object. The higher the frequency (that is, the shorter the wavelength or the bluer the light) the more energetic these escaping photoelectrons would become. Yet, according to standard electromagnetic theory, these electrons' energy should have depended on the intensity (or energy) of the incident light, not their frequency.

As Einstein recalled almost two decades later: "This is highly paradoxical and seems to be incompatible with the fundamental idea of the wave theory." He illustrated this paradox by asking the reader to imagine huge waves generated somewhere on the open sea spreading out in all directions from the center of excitation. Naturally, the wave peaks will attenuate as they propagate farther from the center of agitation. Suppose identical ships are distributed over the ocean. What happens when the waves start to mount? The ships close to the source will keel over or be smashed, but no harm will come to those sufficiently far away—they will just rock harmlessly. One would think that molecules struck by radiation would respond analogously to these ships, yet this is precisely what experience does *not* confirm. Enter the light quantum: "Notwithstanding all due respect for the wave theory, a working hypothesis gained ground that radiation behaves in energetic respects as if it were composed of energy projectiles whose energy magnitude depends only on the radiation's frequency (color) and is proportional to it."[5] According to Planck's law $E = h\nu$, any light quantum (that is, a particle of light) that has an energy above a threshold frequency will have sufficient energy to eject a single electron, creating the photoelectric effect.

His work on the photoelectric effect eventually earned Einstein the 1921 Nobel Prize in Physics, awarded in October 1922. Nonetheless, despite the passage of seventeen years, most physicists remained fundamentally skeptical of Einstein's quantum theory of light. This is largely unsurprising. Einstein had opened his original paper with a bold statement of fact: that there was a fundamental formal difference between the physicists' theoretical conceptions of gases and other ponderable bodies on the one hand—which are treated as countable, discrete bodies—and Maxwell's theory of electromagnetic processes on the other—which was based on continuous fields spread through space. The electromagnetic equations, Einstein said, were insufficient for the explanation of discontinuous phenomena. This incongruity would govern Einstein's work until the end of his life and would animate his later engagement with what would come to be known, after 1925, as quantum mechanics.

Twenty years earlier, Einstein's wonderful year was just getting started. Barely a month after sending off the paper on the photo-electric effect, Einstein submitted his (remarkably short) dissertation to the University of Zurich, entitled "A New Determination of Molecular Dimensions," which similarly solidified the foundations of the atomic theory. The notion that all substances are composed of fundamental particles is of ancient Greek origin, but it lingered for over two millennia in the shadows of more dominant theories that abjured the vacuum. Over the nineteenth century, physicists and chemists regularly began to incorporate indivisible atoms into their reasoning about the microworld, first as a hypothesis and then increasingly as a realistic depiction. By 1905, atomism was ascendant, accepted by most physical scientists except for an influential group of French researchers who argued against it as excessively speculative. Einstein's dissertation computed both the viscosity and diffusion of dilute sugar solutions to get an estimate of molecular sizes, a strategy he expanded on with precision in his December submission "On the Theory of Brownian Motion." The French experimentalist Jean Perrin immediately began testing Einstein's theory, and Perrin's definitive work of confirmation earned him the Nobel Prize in Physics in 1926.

On June 30, 1905—three and a half months after his submission on the photoelectric effect—Einstein sent the *Annalen* what would become his most celebrated paper, "On the Electrodynamics of Moving Bodies." Starting with some fundamental puzzles drawn from the classical theory of electricity and magnetism, Einstein strictly redefined the concept of simultaneity, which in turn generated a new theory of both space and time. Not only did what would eventually be termed the special theory of relativity for uniformly moving bodies make important predictions about the behavior of the electron—a major stumbling block in contemporary physics—but it also in passing dismissed the existence of a luminiferous ether, the putative substrate for light waves thought to permeate the entire universe. A decade later, Einstein would generalize these precepts to accelerated motion, producing a full-blown replacement for Newtonian gravity.

A secondary implication of special relativity—submitted in late September to draw out a conclusion that he had neglected to make explicit in the June paper—was entitled "Does the Inertia of a Body Depend upon Its Energy Content?" As he described it in another letter to Habicht, on September 22, "A consequence of the study on electrodynamics did cross my mind. Namely, the relativity principle, in association with Maxwell's fundamental equations, requires that the mass be a direct measure of the energy contained in a body; light carries mass with it." This was not merely a theoretical construct but entailed the implication that a "noticeable reduction of mass would have to take place in the case of radium. The consideration is amusing and seductive; but for all I know, God Almighty (*der Herrgott*) might be laughing at the whole matter and might have been leading me around by the nose."[6] Over the course of three pages, Einstein derived the equation $E = mc^2$.

Not bad for a Patent Officer, second class. After this string of achievements, Einstein applied to the University of Bern for permission to teach. This petition was denied on the grounds that he had not filed a second dissertation (*Habilitation*) required for the position. Given that any of the *Annalen* papers would have more than sufficed to demonstrate original research, Einstein took the rejection with dark humor and got to work. He submitted his *Habilitation* in early January 1908 and gained a position as a *Privatdozent* (adjunct instructor) at the University of Bern by the end of February. In the summer and fall semesters, he taught courses on the molecular theory of heat and the theory of radiation.

In May 1909, he was appointed as untenured professor of theoretical physics at the University of Zurich for an annual salary of 4,500 francs, the same amount he had been earning at the Patent Office. (This came to roughly half the annual salary of a professor in the United States at that time.) In early July, just before receiving his first honorary doctorate at the University of Geneva, he submitted his resignation first to the Patent Office and then to the University of Bern a month later. On October 15, he left Bern behind. Two weeks earlier, chemist Wilhelm Ostwald was the first person to nominate Einstein for the Nobel Prize in Physics.

112 Mercer Street, Princeton

The Einstein who arrived at New York Harbor on the steamship *Westmoreland* on October 17, 1933 had far different cares from the Bern patent clerk. The intervening years, during which Einstein had lived in Berlin, were especially tumultuous both for Germany and for theoretical physics, with the economic shocks of postwar hyperinflation and the rise of the National Socialist German Workers (Nazi) Party on the one hand, and the headlong development of quantum mechanics on the other. Einstein was a lightning rod of opposition for the Nazis, rendering his time in Berlin increasingly fraught. Meanwhile, his own scientific interests evolved in directions different from the mainstream of the physics community. Leaving Europe made sense.

Princeton was not his only option. Einstein had long entertained offers in Europe—including visiting positions in Leiden and Oxford—but for quite some time he had directed his attention across the Atlantic. He had received his first invitation from an American institution, Columbia University, in 1912. For the two years prior to his final arrival on the *Westmoreland*, Einstein had spent several semesters shuttling between Berlin and the California Institute of Technology (Caltech) in Pasadena. As Hitler's movement gained power in Germany, Einstein made more permanent plans to settle in the United States. During his last semester at Caltech, he decided at first to split his time between Oxford and a new research institution being established under the directorship of Abraham Flexner in Princeton, the Institute for Advanced Study (IAS). When he returned to Europe, however—to Belgium, not Germany—from April to August 1933, he realized that Princeton would need to be his sole home. This was an anxious period, involving a trip to Switzerland in the summer to visit Mileva Marić and his younger son Eduard for what turned out to be the last time. He also traveled to England for various engagements before embarking for the United States.

Although Flexner recognized the triumph of winning Einstein for Princeton, the fact that the physicist was a publicity magnet strained their relationship from the start. Anxious that Einstein should not make any comments to the press when he disembarked, Flexner sent two IAS trustees to whisk him off the ship as it approached harbor and shuttle him to central New Jersey. Flexner monitored Einstein's mail—at one point intercepting a dinner invitation to the White House from Franklin and Eleanor Roosevelt—and continuously irritated the scientist in ways both large and small. Nonetheless, Einstein enjoyed generous financial compensation and the intellectual companionship of brilliant colleagues such as John von Neumann and Hermann Weyl, as well as distinguished visitors such as Paul Dirac and Wolfgang Pauli. The IAS trustees grew increasingly frustrated with Flexner and voted to replace him with long-standing trustee Frank Aydelotte. Einstein was one of three emissaries sent to communicate the bad news to Flexner. Aydelotte held the position until 1947, when he was succeeded by J. Robert Oppenheimer, the wartime director of Los Alamos.

Sometime during the 1930s, Einstein decided to stay in America for good. Having been itinerant and uncertain for so long, he could not have assumed initially that Princeton would be his final perch. There were the problems with Flexner, the uncertainties of exile more generally, and his own quest for stability and quiet. He rented a house at 2 Library Place in Princeton, a relatively short walk from the office. In April 1934 Einstein announced that he would remain indefinitely in Princeton, accepting full-time status at the IAS. In August 1935 he bought a house at 112 Mercer Street, across the road from the rental. He retired from the IAS in April 1945 but continued his frequent walks to the office with mathematician Kurt Gödel for years afterward.

Einstein did not arrive in Princeton alone. The most important companion was his wife (and cousin) Elsa, with whom he had begun a relationship in 1912 and married in 1919, after his divorce from Mileva was finalized. Elsa Einstein had two grown daughters from her first marriage to Max Löwenthal: Ilse (born 1897) and Margot (born 1899). Ilse stayed in Europe when the others departed, dying of tuberculosis in 1934; Margot lived at Mercer Street until her death in 1986. The group was accompanied by Einstein's secretary, Helen Dukas. Einstein's younger sister, Maja Winteler-Einstein, fled her home in Colonnata, Italy, after Benito Mussolini imposed antisemitic laws, joining her brother in February 1939.

The house on Mercer Street was only one sign of Einstein making America his home; citizenship was another. In May 1935, Einstein, Elsa, Margot, and Dukas traveled aboard the *Queen Mary* to Bermuda to apply for American citizenship, to satisfy the requirement that he submit a petition from outside the country. On October 1, 1940, in Trenton, New Jersey, he was sworn in with 88 others, including Margot and Dukas.

Elsa was not with them. After a painful illness, she had died in Princeton in December 1936, leaving the physicist devastated. For more than twenty years she had been his closest companion, nursing him through several ailments, managing his social engagements, and drawing him into various political causes such as pacifism. Einstein's sister Maja died in June 1951, leaving the scientist with Margot and Dukas. Accompanied occasionally by friends, he sailed on the local lake whenever he could.

When Einstein died in April 1955, he was probably the most well-known person in America, a position he had occupied—it would be misleading to say he "enjoyed" it—since his arrival in 1933.

His opinion was sought by journalists and ordinary individuals on almost every issue of importance. As the most prominent refugee during the greatest refugee crisis of the twentieth century, his mailbox overflowed with letters addressed simply to "Albert Einstein, Princeton, USA." Hundreds of family members, acquaintances, and strangers implored him for help with affidavits and visa applications for the thousands upon thousands of displaced European Jews.

A Final Theory?

Einstein's scientific activity in Princeton was both more collaborative and more isolated than it had been in Bern. The collaborative aspect jars with the typical depiction of Einstein as a lone genius breaking the boundaries of classical physics on the strength of his own inspiration. Nonetheless, in the 1930s, following the understandably fallow transitional period of 1932–1933, almost every piece of his scientific work was co-authored. Both the partners he chose and the topics they worked on fell into a specific pattern.

His first collaborator, with whom he co-published seven papers between 1930 and 1934, was the mathematician Walther Mayer, from Graz, Austria. Mayer had had trouble securing academic employment in Central Europe because he was Jewish. In 1929, while Einstein was still in Berlin, Mayer began working with him, following Einstein to Caltech and the IAS as an assistant and continuing to work in Princeton until he died in 1948.

In 1935 *Physical Review* published the most sensational of Einstein's American papers, the so-called EPR (Einstein–Podolsky–Rosen) paper written with Boris Podolsky and Nathan Rosen, entitled "Can Quantum-Mechanical Description of Physical Reality Be Considered Complete?" This paper set out a noticeable challenge to what is often called the "Copenhagen interpretation" of quantum mechanics—associated most prominently with Niels Bohr—by articulating a physical situation that, according to the formalism of the theory, would lead to the conclusion that quantum mechanics is incomplete according to a definition of completeness supplied in the paper. (Both before and after this publication, Einstein never disputed the correctness of the results quantum mechanics provided.) The article has continued, after almost a century, to be the focus of much controversy, discussion, and elaboration. Einstein has been the center of much of that discussion, with his co-authors rarely identified in detail. Podolsky was a Russian émigré who arrived in the United States in 1913. Immediately before collaborating with Einstein, he had worked in Soviet Ukraine. Nathan Rosen was born in Brooklyn and co-authored papers with Einstein on the two-body problem and gravitational lensing in 1936, and then a paper predicting the existence of gravitational waves in 1937.

In 1938, Einstein co-authored a popular-science text called *The Evolution of Physics* with Leopold Infeld, born into a Polish Jewish family in Kraków in 1898. Infeld spent the period from 1933 to 1950 in the United States and Canada before returning to then Communist Poland. In 1938, Einstein and Infeld, together with British mathematician Banesh Hoffmann (also of Polish origin), a student of Princeton mathematician Oswald Veblen, derived equations to describe the motion of binary stars. That same year Einstein wrote a paper on field theory with Peter Bergmann, born in Berlin, who had accepted a position as Einstein's research assistant after completing his PhD in physics at the German University of Prague in 1936.

All of these collaborators were younger than Einstein, and each one was Jewish; all but Rosen and Hoffmann had fled anti-semitism in Europe. In some cases, such as those of Bergmann and Mayer, the position as Einstein's assistant was the sole reason they received American visas. Einstein had worked with collaborators for decades. When faced with mathematical challenges in expressing his physical ideas, he needed the help of colleagues, a practice that reached back at least to Marcel Grossmann at the ETH in 1912 and was continued by this cohort in the 1930s. A new requirement emerged with his emigration: Einstein now published exclusively in English, and his own capacities in the language were limited. He needed bilingual collaborators, such as Rosen, who could transcribe and translate the work.

But it was not just the choice of collaborators, motivated by a mixture of a need for specific skills and a wish to provide succor to marginalized scholars, that characterized the nature of Einstein's work while he was in Princeton—the very topics he was working on were idiosyncratic to him, much to the detriment of his reputation among his contemporaries. The core of Einstein's research from the late 1920s until his death became known as unified field theory: an effort to meld gravity and electromagnetism into a single set of equations. In addition to the aesthetic appeal of unification, one of Einstein's hopes was also to resolve the paradoxes of quantum mechanics. It was a project of extraordinary ambition but he met with little success. Later physicists viewed his work in this period as fruitless and saw the man as being out of touch with the younger scientists of his time.

That criticism was familiar to Einstein. He had started this project in Europe, but he had also continued working on the hot topics of the day such as quantum theory, so his peers could ignore his fixation on higher unification. In Princeton, however, unified field theory emerged as his chief focus and the critics grew louder. From the point of view of today's physics, Einstein's approach was wrongheaded because he did not account for what are today known as the "strong force" that keeps atomic nuclei bound together and the "weak force" that accounts for neutron decay. Presumably, he did not address these forces since he assumed that his unified field theory would pick them up along with all the quantum effects. The public still imagined Einstein was on the path to solving all the problems of physics, as marked by a front-page article in the *New York Times* of February 15, 1950, entitled "Einstein Publishes His 'Master Theory.'" Shortly afterward, Einstein discarded that particular approach.

This older Einstein and his younger self appear to have little in common: the younger had the acumen to lead the way in physical research, while his final years were studded with seeming dead ends in one research project after another. Yet despite the apparent contrasts, there are persistent continuities. In terms of his personal life, one sees consistencies—the challenges of marriage, loss of children, and death; the difficulties of balancing home life and work; and the struggle to care for a growing group of dependents on an income that never seemed to stretch quite far enough—that show how much Einstein was both dependent on, and a support for, others. The same can be said for his scientific partnerships. It is true that Einstein published his 1905 papers alone, yet the tight social network of the Olympia Academy, and especially the companionship of Michele Besso, provided Einstein a rich array of interlocutors to work out his ideas. In that sense, Bern and Princeton were not so distinct.

Einstein's core research principles likewise resonated across the decades. That unified field theory failed should not obscure its affinity with his earlier, highly successful, work. In essence, his efforts at Princeton were an extension of one of the paths he had blazed in 1905 with special relativity and then in 1915 with the generalized theory of gravity. Two main areas of inquiry—the quest for unification and an abiding interest in statistical phenomena and quantum theory—characterized his life's work.

Chapter 2

Relativity Theory

No scientific idea is more closely attached to Albert Einstein than relativity theory, and with good reason. His special theory of relativity, a product of that remarkably creative year 1905, was one of the two chief achievements that established his reputation among theoretical physicists. (The other was his quantum explanation of the photoelectric effect.) Special relativity's successor, the general theory of relativity of 1915–1916—the most substantial advance in understanding gravity since Sir Isaac Newton's *Principia Mathematica* of 1687—made astonishing astronomical predictions that catapulted Einstein to international celebrity in 1919 and continues to make headlines into the twenty-first century with its implications, such as gravitational waves, recognized with a Nobel Prize in Physics in 2017. No scientific idea has made Einstein more intimidating to the uninitiated. The vocabulary associated with relativity is bewildering: curved spacetime, length contraction, energy–mass equivalence, time dilation, gravitational lensing. . . . The individual words make sense, but their combination eludes everyday intuitions.

Einstein had to face the puzzlement induced by relativity theory throughout his life, starting with the term itself. His ideas were first dubbed *Relativtheorie* in 1906 by his supporter Max Planck, the dean of German theoretical physicists, from his perch in Berlin.[1] The term highlighted the theory's rejection of absolute motion and its focus on velocities relative to the observer's frame of reference. Although Einstein also used the term in print in 1907, he did not care for it. He seemed to prefer mathematician Felix Klein's 1910 alternative *Invariantentheorie*—theory of invariance—emphasizing that for him the theory's principal significance was not how measurements of space and time were relative to individual frames, but rather that the theory featured a new way to measure those properties which were invariant across them.[2] The terminology never caught on: relativity theory it is.

Einstein pushed back not only against the popular misunderstanding of the theory that "everything is relative" (it most certainly is not) but also against the myth of its incomprehensibility. He believed that relativity theory could and should be understood by everyone, at least qualitatively. For decades following its formulation, he made repeated efforts to explain special relativity to lay audiences. He did the same for his theory of gravitation, working hard to produce in 1917—on his own initiative—a popular book entitled *Relativity: The Special and the General Theory*. As he wrote to his friend Michele Besso in 1916, after lamenting the pains he was taking: "But if I don't do it, the theory will not be understood, [though] it is now in essence so simple."[3] Readers responded well: fourteen editions appeared within five years, and it was quickly translated into English, French, Russian, Czech, and more.

Einstein's explanations of relativity are best understood through his words, his thought experiments, his analogies. They explain these profound and fundamental scientific theories, and allow us to glimpse Einstein as a teacher. In the act of explaining, Einstein articulated some of the values that animated his scientific research: simplicity, elegance, symmetry. The quest for unification that motivated relativity constituted a central theme of his scientific thought.

Simultaneity, Space, and Time

The basic notions that underlie special relativity had been available for quite some time before Einstein assembled them in 1905. That is not to say that the synthesis was obvious; eminent theorists of his day, including Hendrik Lorentz and Henri Poincaré, failed to push as far or as rigorously as the Swiss patent clerk.

The oldest building block of special relativity was the notion of "relativity" itself. It dates at least to 1632 in *The Dialogue Concerning the Two Chief World Systems*, the book that got the celebrated Italian mathematician and natural philosopher Galileo Galilei into so much hot water. Galileo used relativity to explain why we do not feel the motion of Earth spinning under our feet:

> Shut yourself up with some friend in the main cabin below decks on some large ship, and have with you there some flies, butterflies, and other small flying animals. Have a large bowl of water with some fish in it; hang up a bottle that empties drop by drop into a wide vessel beneath it. With the ship standing still, observe carefully how the little animals fly with equal speed to all sides of the cabin. The fish swim indifferently in all directions; the drops fall into the vessel beneath; and, in throwing something to your friend, you need throw it no more strongly in one direction than another, the distances being equal; jumping with your feet together, you pass equal spaces in every direction. When you have observed all these things carefully (though there is no doubt that when the ship is standing still everything must happen in this way), have the ship proceed with any speed you like, so long as the motion is uniform and not fluctuating this way and that. You will discover not the least change in all the effects named, nor could you tell from any of them whether the ship was moving or standing still.[4]

This phenomenon was so foundational to mechanics that Einstein did not even mention it in "The Electrodynamics of Moving

Bodies," his article in the *Annalen der Physik* that put forth his 1905 theory. (He would, in 1953, contribute an admiring foreword to the English translation of Galileo's book: "The *leitmotif* which I recognize in Galileo's work is the passionate fight against any kind of dogma based on authority.")[5] What Einstein did was extend Galileo's principle from mechanics to electromagnetism.

Where Galileo used a boat, a fitting means of transport for the Italian Renaissance, Einstein chose the signal vehicle of his own age in an article in the newspaper *Vossische Zeitung* in 1914: "We sit in a railway carriage and see (on the adjacent track) another carriage pass by."[6] Once you and Einstein "ignore the vibration of our carriage," you realize that there is no way to determine whether you are stationary and the other carriage is moving, or vice versa. The same inability to determine motion "in reality" also obtains for the other carriage. Although we cannot tell which train is moving (perhaps both are!), the passengers can agree on their *relative* motion. We can seal the carriage and perform many experiments "with all kinds of apparatus imaginable," and the experiments "will come out exactly the same, whether the carriage were not moving or, for that matter, if it were moving at a different velocity." So far, so Galilean.

The problem, Einstein told readers of the *Vossische Zeitung*, concerned light. Physicists had determined through countless optical experiments that "light in empty space always propagates with the same velocity, irrespective of the state of motion of the light source." Moreover, nothing ever travels faster than light: it is a universal speed limit. Light's behavior seems to pose a challenge to the principle of relativity. If the passenger in the carriage across the tracks sees our flashlight emitting light traveling at speed c—this is the standard symbol for the speed of light, although in 1905 Einstein used V—shouldn't she measure that light as going c plus or minus the relative speed of our wagons, the same way she would if we were putting golf balls? Yet she measures c with the same value as we do. "This is where the theory of relativity comes in," Einstein tells us. "This theory shows that the law of constancy of light propagation in vacuum can be satisfied simultaneously for two observers in relative motion to each other, such that one and the same beam of light shows the same velocity to both."

Now things get weird for those of us who generally travel around at speeds far less than the terrifically speedy velocity of light (299,792,458 meters per second, or roughly 671,000,000 miles per hour). We ordinary observers are used to measuring length and velocity without accounting for how long it takes signals from different distant events to reach us. Those events appear to us as simultaneous. In our astronomical age, we are familiar with the idea that the light of the stars was emitted a long time ago and so we are viewing the past when we gaze into the heavens; special relativity reminds us that we are *always* looking at the past because every signal takes time to wend its way to our sensorium. The point is deeper still: even after one accounts for the obvious differences in transit time—say, between the speed of sound and the speed of light after a lightning strike—observers in motion will still disagree about the simultaneity of distant events. "It turns out," Einstein continued in his article, "that two events which are simultaneous with respect to one observer are, in general, not simultaneous with respect to a second observer who is moving relative to the first one. This requires a fundamental change in our concept of time."

Einstein did not include an equation in this 1914 newspaper article, but he did in his 1917 popular book,[7] and at the risk of alienating our own readers slightly it is worth looking at it closely:

$$t' = t / \sqrt{1 - (v^2 / c^2)}$$

This equation states that for each second t that is registered on a clock that is stationary in one frame of reference that is not accelerating, an observer in a different reference frame moving at velocity v with respect to that one will see that clock registering time t'.

This formula, which can be derived in a few minutes with mathematics scarcely more sophisticated than the Pythagorean theorem, leads to astonishing results. We notice that if the two frames are stationary with respect to each other ($v = 0$), then $t' = t$, meaning that the clock registers the same time for both observers. That's a relief: this is what we expect without relativity. As v gets closer and closer to the universal speed limit, c, the denominator of this fraction gets smaller and smaller, which means the measured time t' gets bigger and bigger until it reaches an infinitely large number. (Infinities like this are not physically real, which further confirms that matter will never go that fast.) Therefore, we observe that clocks moving with respect to us run more slowly than clocks in our own reference frame, a phenomenon known as time dilation. This sounds fantastical, but it has been confirmed on countless occasions. One interesting example—unknown to Einstein in 1905—concerns the subatomic particle called the muon, which is generated by radiation hitting our upper atmosphere and then hurtling down to Earth. In their rest frame, muons decay into other particles in about two millionths of a second. However, they are moving at such rapid speeds relative to us that we observe them to decay much more slowly, and we are thus able to measure them reaching Earth's surface.

Einstein derived more consequences from special relativity, including the fact that lengths are observed to contract at very fast speeds in the same way that time expands. (They do not actually shrink in their own frame; it is again a divergence of measurement.) In addition to these counterintuitive but mathematically and empirically unquestionable results, we end up with a fully consistent physics of electromagnetism. Indeed, the constancy of the speed of light implies that Maxwell's equations (formulated in 1865) are what physicists call "relativistically invariant": the same regardless of the state of motion of the observer. This was what Einstein had set out to prove in his original 1905 article, "The Electrodynamics of Moving Bodies," an article that starts with a childlike thought experiment about magnetic induction of a current in a moving coil.

The Superfluous Ether

At the end of the *Vossische Zeitung* article, Einstein included two "major results" of the theory of relativity that he thought were "also of interest to the layman." The first was that "the hypothesis of the existence of a space-filling medium for light propagation, the so-called light-ether, must be abandoned." For most of a century, the existence of such an ether was the bedrock of physical science. Physicists held that light behaved like waves and that waves must travel in some medium—sound waves in air, ocean waves in water—so there had to be a medium wherever light propagated, including the interstellar cosmos.

The ether solved the problem of what light waves were waves *of*, but it created other problems in turn. One of the reasons why electromagnetism had been considered incompatible with the Galilean principle of relativity before Einstein's intervention was that the ether provided an absolute frame of reference. That is, our passengers in their train carriages could discern who was truly moving and who was stationary because both could compare their positions to the ether's. This assumed, of course, that experimentalists and astronomers could detect empirical evidence of the existence of the ether; as of the dawn of the twentieth century, they could not.

The most precise attempt was conducted in Cleveland, Ohio, in 1887 by Albert W. Michelson, who in 1907 became the first American to win a Nobel Prize in a science, in part for this work, and Edward W. Morley. They set up what is now known as an interferometer, an instrument that was greatly expanded by the Laser Interferometric Gravitational-Wave Observatory (LIGO) to detect gravitational waves in 2015. Both LIGO and general relativity lay in the future, though, when Michelson and Morley set to work. (Einstein was then eight years old.)

They erected two axes joined at a right angle by a two-way mirror, so that light could pass through and be partially reflected. The half-mirror would split an incoming light beam, sending two rays racing down the perpendicular legs of the interferometer, which would then bounce off regular mirrors at the end of those legs and return to the half-mirror, recombine, and careen back to the starting point, where a detector awaited them. By having the light beam interfere with itself, the device should detect minute motions with respect to the ether; by rotating the interferometer, one would determine whether the ether was flowing in one direction rather than another. Since Earth was imagined as hurtling through the ether at a tremendous speed, in one direction the light should be pushed along by the ether and in the other direction it should be held up by it, much like a fish swimming with or against a river current. With a device accurate down to a billionth of a meter, Michelson and Morley fully expected to detect this "ether wind." They found nothing.

It is unclear whether Einstein knew of the Michelson–Morley result when he wrote his 1905 paper; he did not cite it—then again, he cited no experiments in the whole paper beyond the simple magnet-coil thought experiment. This was one reason why Einstein's paper looked so anomalous to contemporaries. The only individual referenced was Besso, "for many a valuable suggestion."[8] Einstein was certainly aware of the failure to register empirical traces of the ether. Tacitly building on the approach developed by the Austrian philosopher of science Ernst Mach, whom Einstein had read with his friends in Bern, he declared the assumption of the ether to be "superfluous." It might exist, or it might not, but it had no place in physical theory if it did not leave a trace in measurements.

Einstein's audacious rejection of this foundation of electromagnetism has struck many later readers as his most radical innovation of 1905. It is not obvious that contemporaries took it that way. They continued to publish studies in ether physics for at least a decade after Einstein's initial paper, and experimentalists assimilated Einstein's theory with that of his mentor Hendrik Lorentz—who continued to postulate an ether—without noticing any inconsistency. Slowly but surely, however, the ether began its descent into nonexistence.

It was not entirely dead, however, not even for Einstein. In a lecture for the general public he gave at Lorentz's university in Leiden in late October 1920, entitled "Ether and the Theory of Relativity," Einstein conceded that one could think of "spacetime," a central concept in general relativity, as endowing the ether with some relevance: "More careful reflection teaches us, however, that the special theory of relativity does not compel us to deny ether. We may assume the existence of an ether; only we must give up ascribing a definite state of motion to it [...]." In this framework, "ether" now stood not for a substrate for light waves, nor for an absolute reference frame, but simply for "a medium which is itself devoid of *all* mechanical and kinematical properties, but helps to determine mechanical (and electromagnetic) events."[9]

Such an ether, however, was not what its diehard proponents wanted. Even beyond the antisemitic "Anti-Einstein League" that accused relativity of destroying science, less rabidly reactionary figures occasionally called for a return to the concept. In 1925 experimentalist Dayton C. Miller, working at the Case School in Cleveland, where the original Michelson–Morley experiments had taken place, claimed that his thousands of measurements indicated an ether might in fact exist. When Einstein was asked about Miller's earlier work in 1921, he had responded that if Miller's data were confirmed "then the special relativity theory, and with it the general theory in its present form, fails. Experiment is the supreme judge."[10] In the wake of Miller's updated pronouncements, and six years after his concessions in Leiden, Einstein wrote again in the *Vossische Zeitung* that he was confident that the American's claims would come to naught. Proposing that if the reader "wanted to use this interesting scientific situation to make a bet, I recommend that you bet that Miller's experiments will prove faulty, or that his results have nothing to do with an 'ether wind.' I myself would be quite happy to put my money on that."[11] He would have cashed in: Miller's claims were later dismissed.

The Most Famous Equation

Physicists began drawing out implications from special relativity as soon as it was published, and they have never stopped. The first to do so was Einstein himself, who a few months after the submission of "On the Electrodynamics of Moving Bodies" published a three-page addendum in *Annalen der Physik* entitled "Does the Inertia of a Body Depend upon Its Energy Content?" On its last page, Einstein wrote:

> If a body releases the energy L in the form of radiation, its mass decreases by L/V^2. Since obviously here it is inessential that the energy withdrawn from the body happens to turn into energy of radiation rather than into some other kind of energy, we are led to the more general conclusion:
>
> The mass of a body is a measure of its energy content; if the energy changes by L, the mass changes in the same sense by $L/9*10^{20}$, if the energy is measured in ergs and the mass in grams.
>
> Perhaps it will prove possible to test this theory using bodies whose energy content is variable to a high degree (e.g., salts of radium).[12]

In the *Vossische Zeitung* article of 1914, Einstein framed this point as the second "major result" of special relativity theory: "that the inertia of a body is not absolutely constant, but rather that it grows with the energy content. The important conservation theorems of mass and energy melt into a single theorem: the energy of a body also determines its mass."

Rearranging the algebra—and substituting the now-familiar c for Einstein's V—this is one equation that almost everybody already knows: $E = mc^2$. By 1907, he had given the phenomenon its common name: "the equivalence of mass and energy."[13] For many years, Einstein did not formulate a plain-language exposition of this finding, perhaps because it seemed to him and to many of his colleagues a relatively straightforward consequence of the much more puzzling relativity of simultaneity.

The equation entered his thinking again in 1934, the year after his emigration to the United States. He was reintroduced to his own brainchild by a Princeton graduate student's presentation about the young and dynamic field of nuclear physics—which did not exist in 1905. The student described experiments measuring the effect of changing mass with energy, a formidable challenge given the titanic conversion factor of the square of the speed of light. The physicist Edward Condon, in attendance at the seminar, recalled that Einstein "grinned like a small boy and kept saying over and over, 'Ist das wirklich so?' (Is it really true?)."[14] Einstein was spurred to devise new ways to explain the equation.

His first audience consisted of mathematicians. In the fall of 1934, he delivered the Willard Gibbs Lecture to the American Mathematical Society meeting in Pittsburgh on "An Elementary Derivation of the Theorem Concerning the Equivalence of Mass and Energy." This happened to be Einstein's first scientific lecture given in English, translated for him by his assistant Nathan Rosen. Although apprehensive about his linguistic competence and fortified by prompters lurking nearby, he did not stumble. The address, published soon afterward, began with what he had always considered the signal contribution of special relativity: that it produced an invariant measure of spacetime. Although the Gibbs lecture is formula-studded—no surprise given the venue—the central point was that if "the principles of conservation of impulse [i.e., momentum] and energy are to hold for all coordinate systems" that are moving inertially with respect to each other, then "the presumed equivalence of mass and energy also exists."[15]

Einstein refined this explanation over the years so that it could be digested by a motivated lay audience. He accepted a commission to write an article for the first issue of *Science Illustrated* in April 1946 that he again began with two central conservation laws: those of energy and of mass. Classically, neither allowed for creation or destruction—energy or mass could only change into other forms of energy or mass, respectively. However, the conservation of mass "proved inadequate in the face of special relativity." Einstein skipped quickly to his conclusion: "We might say that the principle of the conservation of energy, having previously swallowed up that of the conservation of heat, now proceeded to swallow that of the conservation of mass—and holds the field alone."[16]

Why does this relationship between energy and mass exist, and what does the speed of light have to do with it? Imagine a reaction between particles happening in a space station, which you glimpse from a spaceship passing by at a very fast speed. Both you and the space station will measure the velocities of the various particles. These will not agree given the transformations for time, length, and velocity derived from special relativity, yet the two sets of readings must obey the conservation of both momentum and energy. The only way to account for the divergence in measurements of energy is to assign the energy as an increase in mass of the particles. That is, one measures particles moving (very fast) with respect to you as being more massive by a small amount; that amount is the energy that has been transmogrified into mass because of the inability of any particle to accelerate beyond the speed of light. Since the result is true in all frames of reference—as all the laws of physics are by the principle of relativity—then it also applies to particles that are stationary (relative to us). Radioactive atoms generate heat when they decay because some of their mass transforms into energy. Einstein tells us by precisely how much.

In *Science Illustrated* Einstein sidestepped the details to focus instead on how hard it was to detect the change in mass, thereby accounting for why the effect had been invisible for so long. He concluded that the release of energy "brings with it a great threat of evil. Averting that threat has become the most urgent problem of our time."[17] After the nuclear destruction of Hiroshima and Nagasaki in August 1945, Einstein's attention, like everyone else's, was focused on the relation's implications for energy. But in 1905, as is clear from his formulation $m = E/c^2$, he was most concerned with how it defined mass. The nature of mass was the central preoccupation of that subsequent decade, leading to arguably the pinnacle of his scientific career.

Gravity

In 1907, the editor of the annual journal *Jahrbuch der Radioaktivität und Elektronik* commissioned from Einstein a review article about relativity theory, necessary because the special theory had begun to attract wider attention among physicists. (Within fifteen years this editor, Johannes Stark, descended into vociferous antisemitism as an early supporter of Adolf Hitler; Einstein became his most loathed target.) The fifty-page article that resulted shows Einstein pushing his ideas in new directions. Naturally, he mentioned $E = mc^2$, a result that he characterized as being of "extraordinary theoretical importance,"[18] but he mostly used that section of the piece as a bridge to his current research: how to generalize the findings of special relativity to apply to accelerated frames of reference.

Whenever he explained to general audiences the resulting theory of general relativity—something he did a great many times after he published the full theory in 1915–1916—Einstein always began with the two principles of special relativity and then added a third. In an unpublished manuscript of early 1920, he described this new idea as having occurred to him while writing the 1907 review. He called it "the most fortunate thought of my life":

In an example worth considering, the gravitational field has a relative existence only in a manner similar to the electric field generated by magneto-electric induction. *Because for an observer in free-fall from the roof of a house there is during the fall*—at least in his immediate vicinity—*no gravitational field*. Namely, if an observer lets go of any bodies, they remain relative to him in a state of rest or uniform motion, independent of their special chemical or physical nature. The observer, therefore, is justified in interpreting his state as being "at rest."[19]

The falling man is correct that all his experimental observations reveal him to be at rest; yet for an external observer, the man and the falling bodies seem to be accelerating in a gravitational field. Yoking together those two concepts would consume Einstein for years.

Just as special relativity began by pondering a mistaken asymmetry between explanations for a moving coil and magnet, the reasoning here also hinged on an asymmetry that lay behind what Einstein called the equivalence principle. This is not the equivalence of mass and energy but rather an equivalence between two definitions of mass. In the classical physics that grew out of the late seventeenth-century brilliance of Sir Isaac Newton, inertial mass was a quantity that measured how hard it was to accelerate an object. In Newton's second law, acceleration is the amount of force applied to an object divided by its inertial mass. Yet mass was also a measure of how much gravitational attraction an object experienced: the attractive force was proportional to gravitational mass. These two concepts of mass occupied different conceptual spaces in classical mechanics, even though every experiment devised had always measured gravitational mass and inertial mass to be exactly the same.

Einstein's "happiest thought" was to wonder how physics would look if we treated them as necessarily the same. This is the lesson of the falling man: the observer sees the effects of gravitational mass, but he who falls could only see those of inertial mass. What if there is no physical difference between accelerating frames and those subject to a gravitational field? In his 1917 popular book, Einstein imagined "a spacious chest resembling a room with an observer inside who is equipped with apparatus" way out in distant space, far from stars or any other body. The observer floats because there is no gravity. But if you yanked a rope hooked onto the chest and accelerated it in one direction, the person "is then standing in the chest exactly the same way as anyone stands in a room of a house on our earth."[20] Einstein limited his example carefully to the special case of uniform acceleration and a homogeneous gravitational field. Later physicists had to work out a more complex issue: Earth (for example) is spherical, so the force projects radially through the floor in a house; the equivalence principle holds only locally. In the space-chest, by contrast, the lines of force are entirely parallel. Our space traveler believes she is experiencing gravity, the inverse of the falling man.

Now imagine a light beam penetrating a hole in the wall of the astronaut's accelerating chest—she would see it bending toward the floor. The light, in traveling a straight path in its own frame of reference, looks curved in an accelerating frame, and therefore also would look curved in a gravitational field. Since light always follows the shortest path, this implies that mass and its equivalent energy bend space and, courtesy of special relativity, also time. Many years later, the Princeton physicist John Wheeler summarized general relativity in a phrase that has stuck: Energy–mass tells spacetime how to curve, and the curve tells energy–mass how to move. Einstein considered the unification of inertia and gravity as one of the most significant achievements of his theory.

The novel mathematics to describe curved spacetime required Einstein to collaborate with mathematicians such as Marcel Grossmann, his university classmate and colleague in Zurich from 1912 to 1914, among others. He learned new geometry and a complex method, the Ricci and Levi-Civita absolute differential calculus—which he called the tensor calculus. He had to worry about how to keep all the conservation laws and principles of special relativity intact. It took many years and several false starts—the "static theory" in Prague, which failed within months; the *Entwurf* ("Outline") theory with Grossmann in 1913, which lasted a little longer; struggling with exotically named paradoxes like the "hole argument"; and so on—but in wartime Berlin in November 1915 he announced his field equations. They remain intact down to the present, comprising one of the most precise and stable theories known to physics.

In his day, the empirically detectable effects that would distinguish his theory from the enormously successful classical model of Newton were vanishingly tiny. Indeed, Einstein described general relativity to the *New York Times* in December 1919 as leading "not to a disavowal of Newton's theory of gravitation, but to a sublimation or supplement of it."[21] Einstein had described three tests of the theory to readers of the *Times of London* a few days earlier. The first he had solved back in 1915: "the distortion of the oval orbits of planets round the sun (confirmed in the case of the planet Mercury)," a minuscule effect that Newtonian approaches had consistently failed to explain. The second—"the deviation of light rays in a gravitational field"—he had since 1911 been coaxing astronomers to investigate by measuring starlight during a solar eclipse. It was finally detected to great fanfare by Arthur Stanley Eddington and colleagues in 1919.

The third test, a "shifting of spectral lines toward the red end of the spectrum" as light approaches Earth from distant stars, known today as "gravitational redshift," interested Einstein and other astronomers almost immediately.[22] It inspired the construction of the modernist Einstein Tower in Potsdam, outside Berlin, which housed a giant interferometer buried deep underground. The effect was extremely hard to disentangle from other solar light effects and was studied extensively by astronomers such as Charles St. John at Mount Wilson Observatory above Pasadena, California. It was eventually first measured successfully by Robert Pound and Glen Rebka at Harvard University in 1960, five years after Einstein's death.

Einstein attended a celebratory conference for Hendrik A. Lorentz, one of his mentors and a pioneer of relativity theory, at the Leiden Observatory in the Netherlands in September 1923. Next to Einstein in the back row are his close friend Paul Ehrenfest and Willem de Sitter, who developed cosmological theories with Einstein; in the front row are Arthur Stanley Eddington, who led the eclipse expedition, and Lorentz himself. *Courtesy AIP Emilio Segrè Visual Archives, Goudsmit Collection.*

The Shape of the Universe

General relativity cemented Einstein's reputation as a great theoretical physicist, but it did not immediately generate a bevy of researchers. Most of his peers were unfamiliar with the abstruse mathematics required to render his equations tractable. Given the challenges of the equations and the minuteness of the effects, most physicists were instead drawn to the fascinating quantum dilemmas posed by the microphysics of the atom. Surveying the publications that elaborated general relativity after the initial flurry of excitement, one detects a demonstrable lull in interest from the mid-1920s to the late 1950s, which one historian has called the "low-water mark of general relativity."[23]

There were, however, notable developments during this period. One came from applying Einstein's field equations to the entire universe, which spawned the new field of physical cosmology. Although the term "Big Bang" to describe the origin of the universe was coined only in 1949—by British astronomer Fred Hoyle, who was belittling the idea, but the name stuck—the question of whether the universe was expanding, contracting, or static occupied several researchers. In 1927, the Belgian theoretical physicist, astronomer, mathematician, and priest Georges Lemaître had been the first scientist to argue for an expanding universe, following upon the 1922 theoretical arguments of the Russian physicist and astronomer Alexander Friedmann. Two years later, astronomer Edwin Hubble published data showing that galaxies were accelerating away from each other, intensifying interest.

Einstein, who was familiar with this work, had early on inserted a particular constant in his field equations, lambda Λ (although initially he used the lower-case form λ), which was dubbed the "cosmological constant." It posited a force that worked against gravity, pushing the universe toward expansion. Einstein had added it to stabilize the universe from a giant crunch lest all matter should attract itself into one lump. He always considered it an artificial term, and his dissatisfaction grew when faced with Hubble's results. As he wrote to Paul Ehrenfest in 1929: "I consider the cosmological matter to be a presumption about which the Lord is merely laughing. The whole theory is not *so* true as to be able to bear such an extrapolation—even though it was fun to consider it just once!"[24] On his trip to Caltech a few weeks later, he publicly acknowledged the work of Hubble, although the worries stayed with him for a long time: "Since I have introduced this term, I had always a bad conscience," he confessed to Lemaître in 1947. "I found it very ugly indeed that the field law of gravitation should be composed of two logically independent terms which are connected by addition. [. . .] I am unable to believe that such an ugly thing should be realized in nature."[25] He spoke too soon. The cosmological constant is now known as "dark energy," accelerating the expansion of the universe, and it constitutes one of the most exciting areas of physics research in the twenty-first century.

Einstein continued to explore the strange astronomical possibilities suggested by general relativity. Greeted at the time with suspicion and sometimes derision, these are now standard aspects of today's astrophysics. Gravitational lensing is one example; gravitational waves are another. In 1916 Einstein suggested that it might be possible for accelerating massive bodies to emit waves in spacetime, in analogy with how accelerating charges produce electromagnetic waves. The effect would be so small and so speculative that nobody addressed it until Einstein and Nathan Rosen returned to it in 1936. In their first effort, they decided that these waves could not exist, and they submitted the article to the American journal *Physical Review*. When a peer-review report suggested corrections, Einstein was so incensed that he withdrew the paper and published a revision in the *Journal of the Franklin Institute*, this time arguing that gravitational waves *could* exist. As of 2015, when it was first measured, we know that Einstein was right. . . the second time.

Einstein was occupied with the development of relativity theory from his first forays into the field in Bern to his final years working on unified field theory. Throughout, he was captivated by the challenges posed by asymmetries in physical theory and how resolving them could yield heretofore hidden unifications. He worked diligently to explain his findings as clearly as possible to a public increasingly fascinated by them. Meanwhile, he was no less engaged with the mysteries of the atom, the dominant area of research for his peers in theoretical physics. There he experienced triumphs much like those with relativity; but unlike relativity, there were also significant reversals.

Chapter 3

Quantum Theory

If the completion of general relativity was Einstein's most strenuous endeavor, by his own account his work in quantum theory was the most revolutionary. The paper on the quantum theory of light was his proudest achievement in 1905, launching the second major strand of scientific contributions which runs through his entire career. Scientists today often think of Einstein as "wrong" on quantum mechanics. One recipient of the 2022 Nobel Prize in Physics exuberantly summarized his own work in the early 1970s as a test of "Einstein's whole platform of doing physics, and we effectively were putting him out of business." Nevertheless to this day, he continued, even he still does not know "what the hell is going on."[1]

Instead of depicting the microworld as a combination of continuously varying quantities, like energy or momentum, quantum theory characterizes it through discrete quantities. An electron does not slide smoothly from one physical state to another, but "jumps" from one to another. Quantum theory scrambles the commonsense perceptions that animate classical physics. Instead, by positing light and electrons as both waves and particles, it insists that complementary physical parameters like position and momentum, or energy and time, cannot be measured simultaneously. Unsurprisingly, such counterintuitive starting points unleash a whole bevy of paradoxes, including apparent contraventions of causality—and yet the theory's achievements in representing physical reality are undeniable.

Einstein was not only intensely involved in the birth of quantum theory and its subsequent elaborations and refinements, he was also engaged in consistently identifying the principal challenges it faced. A proper appreciation of his views requires us to significantly modify the received wisdom on this topic. Therefore, we survey Einstein's ideas about quantum mechanics in the context of the debates he gladly joined, those he was invited into, and those that were thrust upon him. In the 1920s, Einstein was widely recognized and consulted as one of the deepest thinkers about the meaning of quantum theory. Starting in the 1930s, however, his views were forcefully marginalized and often misrepresented, casting him as a physicist antihero who purportedly rejected the scientific community's consensus. This contrast was eventually framed, long after the events in question, as a dialogic contest between Einstein and the Danish physicist Niels Bohr.

It is almost impossible to avoid linking the two together. As it happened, the Swedish Academy decided not to award a Nobel Prize in Physics in 1921, but instead held the prize for that year until 1922, when it gave the delayed prize to Albert Einstein for his 1905 quantum interpretation of the photoelectric effect. The 1922 prize that was awarded that same year was also on a quantum topic: to Bohr for his 1913 model of the hydrogen atom. Both Einstein's and Bohr's work were pillars of what came to be called, after the advent of a fully developed "quantum mechanics" in 1925–1926, the old quantum theory. The question, then and now, is how to understand what happened next, as physicists returned to their work after World War I.

The Compton Effect

By the summer of 1919, when his alma mater, the ETH in Zurich, invited him to speak on quantum theory, Einstein declined, writing that "I don't feel like lecturing on quantum theory. Much as I have labored over it, as little have I succeeded in gaining real insight into it. Besides, I have not sufficiently devoted myself to assembling the many details and tricks of which quantum theory is provisionally composed to enable me to give you all an exhaustive overview."[2] Or, as he wrote to his friend (and quantum physics pioneer) Max Born, "One actually ought to be ashamed of the successes [of quantum theory] because they have been won by the Jesuit axiom: 'Let not thy left hand know what thy right hand doeth.'"[3]

It often helped to discuss thorny physical problems. Paul Ehrenfest kept trying to bring Einstein and Bohr into conversation, triggered by Bohr's own desire to talk to Einstein about his own struggles with understanding the quantum. "I know that no living human has peered so deeply into the veritable abyss of quantum theory as the two of you," Ehrenfest wrote in September 1925, "and that nobody apart from you two really sees how necessary completely radical new concepts are."[4] Einstein wrote back that he would be very happy to get together with Bohr. Indeed, Einstein wrote, there are "mad defenders of principles and there are virtuosos. All three of us"—Einstein, Ehrenfest, and Bohr—"belong among the first sort [. . .]. Hence the effect when encountering utter virtuosos ([Max] Born or [Peter] Debye): discouragement."[5] In the meantime, Einstein caught up with the latest from Hendrik Lorentz's textbooks, from Bohr's papers (often reading them more than once), and from Arnold Sommerfeld's seminal book, *Atombau und Spektrallinien* (*Atomic Structure and Spectral Lines*, 1919). Einstein and Bohr did spend time together at Ehrenfest's house that December, during the fiftieth anniversary celebrations of Lorentz's doctorate. It was during this visit that Ehrenfest took the now-famous photographs of Einstein and Bohr, lounging and smoking against the background

Einstein and Niels Bohr stroll in Brussels during the Fifth Solvay Congress, October 1927. *Courtesy AIP Emilio Segrè Visual Archives, Ehrenfest Collection.*

of an oriental carpet hung on the wall in Ehrenfest's study-cum-living room.

One continuing topic of conflict concerned Bohr's resistance, even in 1921, to the reality of the light quanta Einstein had introduced in 1905. In partial response, Einstein devised several experiments that dealt in one way or another with problems associated with quantum theory. The experiment that excited Einstein the most was one that he was convinced would finally come to a decision between a wave theory or a quantum theory of light. He collaborated on this crucial experiment with Hans W. Geiger, director of the Radium Research Laboratory at the Physikalisch-Technische Reichsanstalt in Berlin, whose eponymous radiation counter would become the ubiquitous tool for the detection of radioactivity. They devised a test that would probe the behavior of light emitted by positively charged ions. They began these experiments in August 1921 and were joined later by Walther Bothe, a member of Geiger's research group. By the end of 1921, Geiger and Bothe reported that no deflection of light was observed, which it should have been if light were a wave. Einstein concluded that the wave field had "no real existence" and that the emission of a quantum was taking place instantaneously.[6] He interpreted these results as a refutation of the field theory of electricity. At this point, Einstein was at his most radical: considering quantum developments, he was prepared to entirely give up the field theory, the base of all physics, including general relativity.

Many colleagues found fault with Einstein's assumptions and reasoning for this test, among them Lorentz and Ehrenfest, who argued that the experiment was not in fact a crucial one. Although Einstein gave a presentation of his initial claims to the Berlin Academy on January 19, 1922, two weeks later he acknowledged defeat due to a calculational error. He once more presented the new derivation of the correct experimental result to the Academy. Einstein often, and quickly, published corrections to his papers in response to criticisms and new developments.

Despite this setback, the collaboration between Geiger and Bothe would produce remarkable results a few years later, when the American physicist Arthur H. Compton conducted an experiment in which he bombarded crystals with X-rays. When an X-ray photon hits an electron belonging to an atom in the crystal, it transfers some of its energy to the electron. The struck atom emits a recoil electron, and this new electron travels off in a well-determined direction while the scattered X-ray, having lost some energy, exhibits a shift in wavelength. Sommerfeld, who was in the United States at the time, wrote to Bohr in January 1923 that if Compton's results were correct, "one would have to drop the wave theory entirely" in the case of X-rays, and "we might be expecting here an entirely new fundamental viewpoint."[7]

In March 1924, Fritz Haber reported to Einstein that, during a visit, Bohr had entertained him for an hour and a half with a "mixture of admiration and disapproval of your light quantum theory [. . .]. He strives with all his being back to the classical world."[8] That spring, together with Hendrik Kramers and John Slater, Bohr published a paper that attempted to refute the light quantum interpretation of the Compton effect—with the consequence of relaxing a strict conservation of energy and hence classical causality. In a letter to Max Born and his wife Hedwig, Einstein wrote: "Bohr's opinion on radiation interests me very much. But I would not like to be driven to abandon strong causality before very different measures have been taken to protect it. I cannot bear the idea that an electron hit by radiation will, of its own accord, choose the moment and the direction at which it will jump away. If so, I would rather be a cobbler or an employee of a casino than a physicist."[9] The Bohr–Kramers–Slater paper was put to a test, again by Geiger and Bothe, who designed an experiment to detect the scattered X-ray quantum and the recoil electron simultaneously. They carried out numerous measurements and lengthy statistical analyses. In the end, their results showed that quantum collisions indeed took place and that there was a statistical dependence between the electron and the scattered light quantum. This was a "triumph for Einstein over Bohr," as Ehrenfest wrote to Einstein.[10]

Einstein began to obsess over the quantum. He considered himself lucky to have other preoccupations "since otherwise the problem of the quantum would have put me into an insane asylum long ago."[11] He saw theoretical physics as being at a "crisis" because of the quantum restrictions on the validity of classical mechanics and electrodynamics. The nature of the theory was still "shrouded in as deep darkness" as it had been fifteen years earlier: "A new mathematical language seems to be necessary; in any case it seems preposterous to express the laws through a combination of differential laws and integral conditions as we do today. Once again, the foundations of theoretical physics are shaken, and experience is calling for the expression of a higher level of lawfulness. When will the saving idea be granted us? Happy are those who may live to see it."[12]

When the Prussian State Library decided to preserve Einstein's voice in a sound recording in 1924, he singled out both the special and the general theory of relativity and the theory of radiation and quanta as the two major areas of his activity. He devoted the brief minutes of this recording to say that he "recognized that mechanics and electrodynamics stood in unresolvable contradiction to observational facts and contributed toward creating that complex of ideas known under the name of quantum theory which has been developed to great fruitfulness especially by Bohr. I shall probably devote the rest of my life to the fundamental clarification of this problem, slight though the prospects of reaching this goal may seem."[13]

He proved quite prescient. Compton's discovery invigorated the community. The years 1923 to 1925 were the high point of quantum publications for Einstein. In April 1924 he wrote an article in the daily press in answer to the challenge: "Does Science Exist for Its Own Sake?" He believed that "every thoughtful person must be given the opportunity to clearly experience the major scientific problems of his day, even when his social position does not permit

him to devote a substantial part of his time and energy to pondering theoretical questions." He proceeded to discuss not the latest
gadget, such as the refrigerator on which he was working, nor the
astronomical discoveries to which he was also contributing, but the
Compton experiment: "Compton's experiment proves that radiation acts as if it were composed of discrete projectiles of energy,
not just as regards energy transfer, but also as regards the collision
effects."[14]

Of Waves and Particles

The following two years were the halcyon days of quantum physics. Even though Einstein published less often on quantum topics,
he remained in intense correspondence with the major players.
In these letters he continued to be critical, but he refrained from
expressing his objections in print. His engagement helped to clar-
ify the two new frameworks for "quantum mechanics": the matrix
mechanics of Werner Heisenberg, Max Born, and Pascual Jordan,
and the wave mechanics of Erwin Schrödinger. Einstein preferred
the latter. He embarked on a long collaboration with a relatively
unknown young physicist, Emil Rupp, on an experiment that
would decide whether excited atoms emit light instantaneously (in
quanta) or in a finite time (in waves). As he had earlier, he expected
quantum emission to be confirmed by experiment. In the midst
of the collaboration, Einstein wrote to Born what later became
his most famous comment on the subject: "Quantum mechanics
is very worthy of respect. But an inner voice tells me this is not
the genuine article after all. The theory delivers much but it hardly
brings us closer to the Old One's secret. In any event, I am convinced that *He* is not playing dice."[15]

Einstein did not participate in the 1927 International Congress
of Physicists on Lake Como at which Bohr provided a prominent
exposition of his evolving philosophy of "complementarity," a

pillar of what would come to be known as the "Copenhagen inter-pretation" of quantum theory—so named because of the location of Bohr's institute, where many of its exponents, such as Werner Heisenberg, had worked out their views. Bohr was concerned with the conditions of observation and description within quantum mechanics. He argued that quantum mechanical phenomena could only be fully explained with complementary descriptions: one would have to employ *both* classical concepts of wave and particle, even though these descriptions were now mutually exclusive. The observer must decide which aspects of the phenomena to focus upon. Once this choice is made the contradictions disappear, because the observer destroys the "possibility of the realization of conflicting aspects."[16] Bohr insisted that physicists could abandon classical physics while not yet fully embracing the light quantum (or "photon").

In this respect, Bohr was a more conservative physicist than Einstein, who had been willing to jettison electromagnetic field theory on several occasions. Bohr struggled to elaborate on the nature of descriptions and concepts used in physics, and thereby opened philosophical deliberations on the foundations of physics. He argued that the human observer of physical events cannot be separated from those events, that "the new situation in physics has so forcibly reminded us of the old truth that we are both onlookers and actors in the great drama of existence."[17]

In other respects, Bohr was more radical. A year later, also in Italy, Bohr condemned Einstein with faint praise to the relics of the past: "Indeed, the whole conceptual structure of classical physics, brought to so wonderful a unification and completion by Einstein's work, rests on the assumption [...] that it is possible to discriminate between the behavior of material objects and the question of their observation. For a parallel to the lesson of atomic theory regarding the limited applicability of such customary idealizations," he wrote, one ought to turn to the Buddha or Lao Tse, a suggestion that he followed up by repeating the melodramatic invocation of "the great drama of existence."[18]

After 1927, Einstein published only short articles on quantum topics. He fell seriously ill in 1928 and was not able to resume intense work until early 1929, when he focused on his major pursuit: unified field theory. That year Einstein, and the rest of the world, celebrated his fiftieth birthday. Afterward, he wrote numerous articles on topics related to pacifism and internationalism, and he traveled often to and from the United States and England. By March 1933, Einstein knew he would never return to Berlin. On June 10, 1933, he gave the Herbert Spencer Lecture at Oxford University "On the Method of Theoretical Physics," his last notable appearance in Europe, in which he articulated much of his philosophy of science, including some criticisms of Bohr's complementarity.

The Entanglement Paradox

In 1935, Einstein returned to these critiques in his last major publication on the quantum, the Einstein–Podolsky–Rosen (EPR) paper. The manuscript was received by *Physical Review* on March 25 and was published on May 15, 1935. During the intervening two months, this paper, like almost everything that Einstein now did, was hyped up in the media. On May 4, 1935, the *New York Times* reprinted a Science Service news item, announcing that Einstein "attacks quantum theory" of "which he was a sort of grandfather": "Scientist and two colleagues find it is not complete, even though correct."[19] Three days later the *New York Times* published a one-paragraph response from Einstein, in which he chastised the newspaper for publishing scientific news before the article was in print. Einstein and Rosen also wrote on May 10 to the director of Science Service, restating that a press report ought never to appear before the work itself and questioning what usefulness such a piece could have for the public. They characterized their article as "of interest only to professional men" and wished to avoid the impression "that authors are seeking to advertise themselves."[20]

The EPR paper presented a criterion for physical reality: "*If, without in any way disturbing a system, we can predict with certainty (i.e., with probability equal to unity) the value of a physical quantity, then there exists an element of physical reality corresponding to this physical quantity.*"[21] The authors then proceeded to argue that quantum mechanics fails this "reality" criterion in the case of a thought experiment involving a pair of particles in what Erwin Schrödinger would shortly thereafter call an "entangled" state. When the particles briefly interact and then move off in opposite directions, Heisenberg's uncertainty principle dictates that it will be impossible to simultaneously measure precisely both the position and the momentum of either particle. According to quantum mechanics, the exact values of either the momentum or the position of particle 2 can however be found by measurements carried out solely on particle 1. According to the EPR paper this meant that, even after their separation, the second particle must simultaneously have exact values of position and momentum. Since quantum theory did not allow for that, something was amiss.

Einstein had been thinking about these ideas for quite a while, at least since the Fifth Solvay Conference in Brussels in October 1927, although all we know is based on the recollections of others, since Einstein made no presentations at the conference. He did not approve "of the purely statistical way of thinking on which the new theories are based," and he was "still looking for a theory that is fully deterministic."[22] Einstein's only comment recorded in the conference proceedings, following Louis De Broglie's presentation, was along the same lines. He imagined an experiment and the two interpretations of such an experiment but found both wanting. At the heart of his argument was a concern with the nonseparability of a quantum mechanical system—that the parts of such a system cannot, in general, be completely specified independently of the whole, even when those parts are widely separated.

The argument immediately generated some discussion, if not exactly a fierce debate. The only contemporary record we have of a vociferous reaction lies in a letter by Ehrenfest a week later to his physicist colleagues Samuel Goudsmit, George Uhlenbeck, and Gerhard Diecke: "Bohr from out of the philosophical smoke clouds constantly searching for tools to crush one example after another. Einstein like a jack-in-the-box: jumping out fresh every morning. Oh, that was priceless."[23] The putative conversations between Einstein and Bohr may have taken place at mealtimes, or during walks, and the textual evidence remains scarce. There is, however, a short (and rather puerile) note that appears to have been passed back and forth between Einstein and Ehrenfest during Compton's report. Ehrenfest wrote to Einstein: "Don't laugh!! There is a special department in purgatory for 'professors of quantum theory,' and there you will receive lectures on classical physics for ten hours a day." Einstein responded: "I'm only laughing at the naïveté. Who knows who'll be laughing in a few years?"[24]

Erwin Schrödinger, for one, was delighted by the EPR argument. He wrote from Oxford to congratulate Einstein for having "publicly grabbed the dogmatic quantum mechanics by the scruff of the neck."[25] Einstein responded quickly, explaining that the paper had been written by Podolsky for language reasons (i.e., in English) after many discussions. Yet it had not come out the way Einstein had wished: "The difficulty is that physics is a kind of metaphysics; physics describes 'reality.' But we do not know what 'reality' is: we know it only through the physical description!" This letter criticizes "Talmudic" thinking. The only way to get around the "Talmudiker" is to have a principle of separability (*Trennungsprinzip*). If you accept this principle, then you get rid of the Schrödinger interpretation of the quantum formalism—namely that a particle cannot be said to be in one of two boxes until you look—and what remains is Max Born's interpretation: the Schrödinger equation describes the probabilities of the ensuing measurement. Yet such a theory would, for Einstein, be *incomplete*. In the penultimate paragraph of this letter, he wrote: "But I must observe that I do not believe that we must be satisfied with an 'incomplete' description of the real state of things, but that we should look for a complete description."[26]

In the meantime, many letters were exchanged among other European colleagues. Wolfgang Pauli wrote a rather angry missive to Heisenberg about his distaste for the EPR argument, voicing dismissive remarks about Einstein's co-authors ("not very good company"). He thought this collaboration was catastrophic since such publications could lead to "confusion in public opinion" in America. Pauli worried about the image of quantum mechanics when "older gentlemen like Laue and Einstein are bandying around the idea that quantum mechanics, while correct, is incomplete."[27] His negative sentiments may have been an echo of an exchange with Einstein over five years earlier. In a haughty critique of Einstein's new approach to unified theory of electromagnetism and gravity based on distant parallelism, Pauli had wagered that Einstein's work would eventually lead to nothing. He saw Einstein's attempts as a "betrayal" of the general relativity theory by "having gone over to join the mathematicians."[28] Einstein was not pleased:

> I by no means assert that the path which I have embarked upon must necessarily be the correct one. But I do assert that it is conceptually the most natural path that I have thus far encountered. Before the mathematical consequences have been properly thought through, it is by no means justified to make negative judgements about it. It is not correct that I repudiate the progress that has been made in quantum theory. I simply believe than one cannot penetrate to the depths of the problem with these semi-empirical methods, and that being satisfied with only statistical laws will later be rejected.[29]

Further news on EPR appeared in summer 1935. *Nature* published a paraphrase of the argument on June 22. In response, Bohr wrote to the journal's editors that he intended to publish a detailed reaction in *Physical Review*, which *Nature* printed on July 13. By then, Heisenberg had written back to Pauli, saying that he had heard about Bohr's forthcoming article. He also included a manuscript of a critique that he himself had written, which was never published.

Bohr's response was received by *Physical Review* on July 13 but was not published until three months later. In the meantime, on July 28, a sensationalist article appeared in the *New York Times*, setting up an Einstein versus Bohr confrontation. Its author was William L. Laurence, an enterprising journalist with a penchant for exaggeration and headline-grabbing, who would write similarly provocative pieces in 1945 on the atomic bomb as the Manhattan Project's embedded journalist.

While Bohr's response was pending, Einstein's private exchanges with Schrödinger continued. The EPR argument had raised the counterintuitive nature of quantum superpositions, in which a quantum system such as an atom or photon can exist as a combination of multiple states corresponding to different possible outcomes. According to quantum wave mechanics, the wave function (ψ) which describes a quantum state "collapses" into one or another of possible states precisely when that state is observed or otherwise disturbed. Addressing this superposition of possible states, Schrödinger now illustrated it for a macroscopic system by introducing his eponymous cat argument in a letter of August 19 to Einstein.

He imagined a cat locked in a box. The cat's life depends on whether an atom of radioactive material has emitted its radiation or not, triggering a mechanism that would smash a vial of poison. (Radioactive decays are entirely random processes.) According to the Copenhagen interpretation, the cat would remain both alive and dead until the moment that the cat and the atom are observed, at which point the wave function collapses into one or another state. With this illustration, Schrödinger wished to satirize the puzzles posed by quantum mechanics but ended up opening a whole new vista of physical and philosophical debates.

No written exchanges between Einstein and Bohr seem to have taken place during this entire time. In July, Einstein was asked by the Danish press to contribute an article to congratulate Bohr on his upcoming fiftieth birthday. Einstein sent a one-page comment, which duly appeared in translation in *Politiken*, the major Danish newspaper. Einstein wrote that "Niels Bohr's brilliant inventiveness" regarding the structure of the atom and nature of the forces that act on atomic transformations "constitute lasting scientific contributions. Without his work, one could not even be able to imagine quantum theory. Bohr stands as a guarantee against the destruction of civilization by the dark forces of confusion and fanaticism."[30]

When Bohr's response to EPR finally appeared in mid-October, its several densely argued pages did not settle the issue. After briefly summarizing EPR's argument, Bohr denied that the authors had uncovered any paradox because quantum mechanics forbids ever measuring one system without also measuring the other; that is, he refused to grant the *Trennungsprinzip*: that one could conduct the EPR experiment "without in any way disturbing a system." Bohr illustrated his point with a different thought experiment of a particle passing through a slit in a diaphragm. This was indeed "a radical revision of our attitude as regards physical reality," but he concluded that it was not necessarily more extreme than the one Einstein himself had caused earlier, namely, "the fundamental modification of all ideas regarding the absolute character of physical phenomena, brought about by the general theory of relativity."[31] Neither Einstein nor his collaborators wrote a rejoinder.

Writing the Winners' History

This "Einstein–Bohr debate" is an artifact. It is the result of a retrospective narrative, memorialized while Einstein was still alive, in a volume of essays dedicated to him on his seventieth birthday in 1949. The project was long in the making. As editor of the *Library of*

Living Philosophers, philosopher Paul Arthur Schilpp had produced volumes on John Dewey, George Santayana, Alfred N. Whitehead, G. E. Moore, and Bertrand Russell, and he now turned to Einstein. The essays in this collection have had an outsized impact on how the physicist has been remembered, in part because of the "Auto-biographical Notes" that Schilpp commissioned—which Einstein called his "obituary" and completed in 1946, three years before its publication and nine years before his death—alongside contributions from numerous scientists and several philosophers. Schilpp also asked Einstein to write responses to his colleagues' essays.

Bohr's article in this volume constitutes the origin of the long-lived narrative that a profound debate with Einstein had taken place over many years. The arguments in this essay, which took Bohr two years to write, were later reinforced through his interviews by historians of quantum theory in the early 1960s, after Einstein's death. Bohr's numerous colleagues, students, and post-docs, immersed in his views and reminiscences for more than half a century, erected a version of the history of events that exaggerates the length and depth of the disagreements between Einstein and Bohr. Einstein himself did not have an equivalent "school" or group of acolytes.

Einstein was curious to read Bohr's article, although he explained beforehand that he knew "Bohr's point of view pretty well . . . opinion opposes opinion, and a convincing decision will have to wait for a long time."[32] He eventually received Bohr's essay in early August 1948. By then Einstein was in poor health. He suffered from severe abdominal pain that was diagnosed as a large aortic aneurysm. He underwent major surgery to prevent a rupture and spent three weeks at Brooklyn Jewish Hospital under the care of Dr. Rudolph Nissen, who carried out an innovative procedure of wrapping Einstein's aneurysm in cellophane, allowing Einstein to reach his seventy-sixth birthday. Einstein sent Schilpp his thirty-four-page "Reply to Critics" in late March 1949, explaining that he had been unable to address essays that had arrived too late. The volume was typeset during the summer, and proofs reached Einstein in early September 1949. He later signed 760 copies of a special edition of the volume, complaining of cramps in his right hand.

Bohr's forty-page contribution, "Discussion with Einstein on Epistemological Problems in Atomic Physics," collapsed what had been a broad conversation about the philosophical implications of quantum mechanics among the community of theoretical physicists in both Europe and North America into a single debate that these two individuals had conducted together. This not only transformed the complicated give-and-take of a many-sided inquiry into a clash of two titans—for Einstein and Bohr were unquestionably the leading theorists of the interwar period—but it also endowed Bohr's position with greater coherence than it had possessed at the time of the EPR paper. The essay produces this effect not by presenting a rigorous philosophical position but rather by offering a historical one, enumerating a series of discussions Bohr had with Einstein beginning with the disagreements that Ehrenfest reported at the Solvay Conference of 1927—setting aside the less adversarial conversation in Ehrenfest's living room in 1925.

Bohr claimed that the exchanges with Einstein "took quite a dramatic turn" at the Solvay meeting of 1930, as Einstein proposed a thought experiment involving a clock and a photon which again invoked the issue of separability, which Bohr countered by invoking an effect from Einstein's general relativity. It was at this moment that Bohr laid down the terms of the history that has become received wisdom among the mainstream of physicists and philosophers ever since:

> Notwithstanding the most suggestive confirmation of the soundness and wide scope of the quantum-mechanical way of description, Einstein nevertheless, in a following conversation with me, expressed a feeling of disquietude as regards the apparent lack of firmly laid down principles for the explanation of nature, in which all could agree. From my viewpoint, however,

I could only answer that, in dealing with the task of bringing order into an entirely new field of experience, we could hardly trust in any accustomed principles, however broad, apart from the demand of avoiding logical inconsistencies and, in this respect, the mathematical formalism of quantum mechanics should surely meet all requirements.

Bohr's statement here does not address the efficacy of quantum mechanics—which Einstein never disputed—nor any physical or philosophical argument against the *Trennungsprinzip*. Instead, Bohr places Einstein on the side of those clinging to "accustomed principles" and declares himself willing to stride into the future without first developing a thorough foundational explanation. What came to be called the "Copenhagen interpretation" was a combination of three heuristics (the correspondence principle, the principle of complementarity, and Heisenberg's uncertainty relation) that did not fully respond to the critique of its completeness. Bohr did report on additional discussions after the publication of EPR, including one visit to Princeton in 1937 "where we did not get beyond a humorous contest concerning which side Spinoza would have taken if he had lived to see the development of our days." Further conversations that took place in the 1940s, up to the moment of writing the article, "have so far not led to a common view about the epistemological problems in atomic physics."[33]

And so the "Bohr–Einstein debate" was born: Bohr put limits on what can ever be known, while Einstein hoped that future work would break through these barriers. It obviously served Bohr's own position to present the argument with Einstein in this way; it is a separate matter as to why this account has been accepted by others as showing Einstein being out of touch with, or even opposing, quantum mechanics. There are two answers to this question, one that addresses the 1940s through the 1960s, and one that began in the 1970s and has reached resolution only in the early twenty-first century.

Case Closed?

Between the publication of the EPR paper and Bohr's summary article about his conversations with Einstein lay fourteen years and the cataclysm of a global war, which fundamentally reshaped the geography and practice of physics. The center of gravity of the discipline had already begun to shift to the United States in the 1930s—Einstein's settling at the Institute for Advanced Study was part of a more general pattern. In the immediate postwar years, the advent of nuclear weapons shone a spotlight on American physics, which experienced enormous growth as a devastated Europe rebuilt its institutions of higher education. The first generation of quantum physicists, including Einstein, had been exposed to Plato and Kant during their early schooling, and they were thus primed to take up epistemological debates. The new American generation had neither the background, the inclination, nor (in overcrowded classrooms) the time to indulge in queries about the deeper significance of the collapse of the wave function. In this atmosphere of "shut up and calculate," foundational queries about quantum mechanics went into hibernation. Hadn't Bohr already resolved those issues by besting Einstein? You could look it up if you were curious, but it wasn't real physics.

Starting in the 1960s, a new generation of physicists began to turn again to the puzzles posed by separability and entanglement in quantum mechanics, what Einstein had called in 1947 in a letter to Born "spooky actions at a distance."[34] The starting points of their efforts was a paper published by John Bell in 1964 entitled "On the Einstein-Podolsky-Rosen Paradox." Opening with the thought experiment of two entangled particles that Einstein and his colleagues had posited almost thirty years earlier, and which Einstein had raised already at the Solvay Conference of 1927, Bell argued that in order for the *Trennungsprinzip* to hold, there would have to exist "hidden variables" that would account for the behavior of the separated particles; such variables would produce empirical consequences that were different from what the formalism of quantum mechanics predicted. In short, Bell had accomplished what Bohr had not: he rendered the EPR argument into a framework that could be experimentally tested. There were only two problems: quantum experimentation of this sort was devilishly hard to do; and nobody cared.

For several years, no one even cited Bell's paper, much less attempted to test it. A graduate student named John Clauser came across the article and proposed the experiment as a dissertation topic, but his advisor scotched the notion and steered him toward something more practical. Degree in hand, Clauser in 1972 published results that disproved the separability argument of the EPR paper by showing that two widely separated particles can in fact be entangled. Again, nobody paid attention. Starting in 1982, a French physicist named Alain Aspect and his team published a series of experimental studies of Bell's theorem, confirming the predictions of quantum mechanics to an astonishingly precise 242 standard deviations. In the 1990s, Austrian physicist Anton Zeilinger took up tests of his own. In 2022, all three were awarded the Nobel Prize in Physics for finally refuting Einstein's critique of the completeness of quantum mechanics—a critique so compelling that it had captivated physicists for most of a century.

There was not really a "debate" between Einstein and Bohr until Bohr created it and then declared himself the winner—prematurely. The physics community had decided that Einstein had lost to Bohr long before the experiments of Aspect, Clauser, and Zeilinger demonstrated that the supposed paradox posed by the EPR thought experiment was actually our reality. The work of these physicists has not put to rest all the philosophical issues raised by Einstein and others about quantum mechanics; for example, debates about the problematic relations between general relativity and quantum mechanics continue to the present. The *story* of the debate and its resolution, however, has functioned as a cautionary tale: even the great Einstein could be wrong about important questions in physics.

Chapter 4

Belonging

Today Einstein has been adopted as a cultural icon by Germany, Switzerland, the United States, and Israel; by the cities of Ulm, Bern, Zurich, Berlin, Leiden, Prague, New York, Princeton, Pasadena, Washington, DC, and others; by various universities; by scientists around the world; by pacifists (for a while); and by those who market educational gimmicks to enhance intelligence and curiosity. Trying to figure out the "correct" box into which we can slot Einstein is a fool's errand. He might be characterized by what we today sometimes call "cosmopolitan" and what he often described as "European." He carried various passports over time, and he abandoned his German official identity twice: the first time before age sixteen, so as not to be eligible for the draft; and a second time almost forty years later, when he became a refugee from Nazi Germany upon Hitler's ascent to power in early 1933. He had elective affinities for some places, such as Switzerland and Italy (where he lived in 1895 and where his family resided much longer), and his experiences in these countries would shape how he looked at the world. Einstein's many peregrinations and his diverse political and humanitarian engagements also shaped how the world looked at him and his various identities, some of which he claimed for himself and others which were imposed upon him from outside.

In almost every context, one encounters Einstein's relationship to the Jewish community. This relationship was no less variable over time than his scientific and political views were. If he has remained a hero to Jews worldwide, it is because his commitment was, at a certain level, elemental. Neither biological nor racial nor cultural, his sense of belonging among Jews was that of an emotional tribal, familial, and historical affinity—a diffuse, nonideological attachment grounded in a perception of shared suffering across generations. This was not a religious position; indeed, he insisted that he was "nonaffiliated" (*konfessionslos*) with any religion. He made this position clear in official documents starting as early as 1896, when he requested release from his Württemberg citizenship. He refused to pay taxes to the Berlin Jewish community even though he was required to do so by German law. Thus, over the years he maintained his stance as a nonreligious, secular Jew—but a Jew nonetheless—willing, and demonstrably able, to devote tremendous effort to various Jewish educational and welfare causes. He refused to aid in campaigns to combat antisemitism, which he viewed as a problem of non-Jews, and yet this very antisemitism helped form his own Jewish identity again and again. As he wrote to the German politician Willy Helpach in October 1929: "When I came to Germany fifteen years ago, I first discovered that I was a Jew, and this discovery was communicated more by non-Jews than by Jews."[1]

That this statement was in a letter is significant. Among his many other commitments and interests, Einstein was a tireless reader and writer of letters. Health permitting, he prioritized responding to his many correspondents, although the volume of his mail became so large in later decades that he had to be more selective. The letters to and from Einstein are the crucial sources for understanding how he consistently worked through questions of belonging. He was perceived throughout his lifetime mostly along three axes: Einstein the German, Einstein the Zionist, and Einstein the American.

To and from Bavaria

A lively, optimistic man, Albert Einstein's father, Hermann Einstein, was born in Buchau, on Lake Feder, in the Kingdom of Württemberg in southern Germany, at a time when its Jewish population had reached its zenith. Jews had first settled there in the fourteenth century. They were mostly peddlers, excluded from the professions and from land ownership until 1828, when they gained a measure of civil rights. The town had its own synagogue, which was inaugurated in 1839 in the presence of King Wilhelm I. Hermann attended a Gymnasium in Stuttgart for only one year, at age fourteen, after which he received release from the compulsory three-year military service. His financial situation did not permit him to continue his education. He worked in a cousin's business in Ulm and later in the featherbedding company Israel und Levi before marrying Pauline Koch in 1876.

Three years later, their eldest son Albert was born. Ulm's registrar recorded on the birth certificate that at half past 11:00 in the morning of March 14, 1879, a male child was born to the merchant Hermann Einstein, residing in Bahnhofstrasse B No. 135, of Israelite religion, and to his wedded wife Pauline née Koch, also of Israelite religion, who lived with him in his Ulm residence. The child was given the first name Albert—not a family name and not a typically Jewish name, but a *German* name, quite popular in this first decade after the unification of the German Empire in 1871.

In June 1880, when Einstein was just a year old, the family moved 120 kilometers east to Munich, the political and cultural center of the southern German state of Bavaria. The Jewish population of Munich was then 4,144 of a total of 230,000 inhabitants. It almost doubled in the subsequent two decades after significant immigration from Eastern Europe following pogroms in Russia and elsewhere, but it never exceeded some 2 percent of the rapidly expanding city population. The Jewish community was strongly assimilationist, with a 24 percent intermarriage rate.

Hermann and his brother Jakob, who had studied at a technical college, settled first on Müllerstrasse and then in a suburban compound on Adlzreiterstraße, about 1.3 kilometers from Blumenstrasse, where young Albert would go to elementary school. The house, located in a large garden in the Sendling suburb, was near where their J. Einstein & Cie. electrical factory stood. The company produced dynamos, lamps, a telephone system, and other electro-technological supplies, some of which were exhibited at the Munich International Electrical Exhibition in 1882. The business thrived, with regular injections of capital from the Koch and Einstein family elders, growing to two hundred employees by 1890. The two brothers also built power and lighting systems first in Munich and later in Italy.

As a young boy, Einstein poses in an artist's studio, Munich, 1880s. *Courtesy Schweizerisches Literaturarchiv: Nachlass Jost Winteler, SLA-Winteler-Akt1.*

According to family lore, at birth, the back of baby Einstein's head was unusually large and angular, but soon thereafter it reverted to a normal shape; he was said to be heavy and quiet, required little care, and was content to play by himself with blocks and puzzles for hours. It also holds that he learned language slowly and that his parents feared he would never learn to speak. He received private instruction at home and then entered the second grade of the Petersschule, a state Catholic primary school, at age six and a half. The classes were exceedingly overcrowded. For example, Einstein's third and fourth grade classes each had seventy students and included children from various economic backgrounds. He was always first in his class. It was at school that Einstein became aware of his socioeconomic class as the child of a solidly bourgeois, comfortable, but not rich family. He studied reading, grammar, calligraphy, singing, and gymnastics, and in addition was taught six hours of arithmetic every week. The only Jewish student in the school, he was required to study the Catholic religion; his father, who prided himself as a free-thinker, ensured nevertheless that Einstein received private Jewish religious education at home in preparation for his bar mitzvah. Einstein was at first quite engrossed in those studies, but his interest abruptly ceased around age twelve, in large part because of his growing fascination with science. In later years he regretted having been a lazy and obstreperous pupil during the study of Hebrew language and literature.

At the age of nine and a half, Einstein entered the Luitpold Gymnasium, which had opened only a year earlier. German humanistic high schools emphasized the study of Greek and Latin, and Einstein received good grades in both; he particularly enjoyed ancient history and German literature. He consistently received the highest marks in mathematics, where he discovered the pleasures of algebra and of mathematical and geometrical proofs. The subject interested him so much that he would open his textbooks during the summer vacations, ahead of the start of a school year. His later memories of school were at best ambivalent, however.

Einstein began to reject the strict, disciplinarian demands of teachers, the adulation of authority and empire, the emphasis on rote learning, and the anxiety of final exams. He was also made acutely aware, at a rather young age, that he belonged to a religious minority. In 1920, in a letter to the political editor of the *Berliner Tageblatt*, Paul Nathan, he commented that in a recent article on antisemitism

> I refrained from describing my own experiences from my school-days in Munich, since these experiences are not particularly meaningful for those not involved. The teaching faculty at elementary school was liberal and did not make any denominational distinctions. Among the secondary school teachers there were a few antisemites, in particular one who flaunted his rank as reserve officer. Among the children antisemitism was alive particularly at elementary school. It was based on the children's remarkable awareness of racial characteristics and on impressions left from religious instruction. Active attacks and verbal abuse on the way to and from school were frequent but usually not all that serious. They sufficed, however, to establish an acute feeling of alienation already in childhood. It is not worth wasting much breath on that in the article.[2]

After Einstein's father and uncle founded the Einstein & Garrone Company in Pavia in March 1894, the family moved to Milan, where the business offices were located. The young Einstein was left in Munich with relatives to complete his last year of the Gymnasium and his final examinations (*Matura*). Those plans fell through in late December 1894 when his homeroom and Greek language teacher, Joseph Degenhart, whose instruction methods Einstein demonstrably disliked, impugned the young man with undermining classroom respect. Einstein sought a medical excuse and withdrew from the school. He left immediately for Milan. It was his first emigration.

He spent the next nine months in Italy, studying physics by himself and writing his first scientific essay, "On the Investigation of the State of the Ether in a Magnetic Field," even though he had not formally studied physics in Munich for more than a few months. Two years under the admission age and without a *Matura*, armed with a recommendation from his former mathematics teacher, Einstein presented himself for an entrance examination to the Eidgenössische Technische Hochschule (ETH) in Zurich in September 1895. He failed some topics in the general knowledge portion of the examination, although he did sufficiently well in the scientific part to be permitted to attend the physics lectures of the prominent physicist H. F. Weber. On the advice of the Polytechnic's director, Einstein enrolled for a year in the Aargau Kantonsschule, a Gymnasium in the town of Aarau, 50 kilometers west of Zurich, where he completed his secondary education with excellent grades in the summer of 1896. This was his second emigration.

The East

Einstein's first years in Berlin, from 1914 to 1918, were some of the most intense and stressful years of his life. He lived in relative isolation, devoting himself to scientific work, while his wife and children, after a brief time in Berlin, returned to Zurich. His correspondence with colleagues was entirely focused on his program to complete the general theory of relativity, achieved in November 1915, in addition to much work on foundational questions in quantum theory during the same period. He was, and wished to be, enfolded by the Einstein family in Berlin, especially his cousin Elsa (who in a few years would become his second wife) and her parents, who cared for Einstein and nursed him during his various bouts of ulcerative colitis and major exhaustion in 1917. Elsa's daughters Margot and Ilse and their circle of friends provided the warmth and familiarity he had not experienced since childhood. But new close friendships

Einstein and his secretary Helen Dukas work in his office, the so-called tower room, above his apartment at Haberlandstrasse 5, Berlin; this image dates from about May 1929. *Courtesy of the Albert Einstein Archives, The Hebrew University of Jerusalem.*

comparable to his small circle of confidants from Switzerland did not develop. He repeatedly resisted attempts to claim him as German, briskly admonishing one well-intentioned correspondent in 1918 that "I am by heritage a Jew, by citizenship a Swiss, and by mentality a human being, and only a human being, without any special attachment to any state or national entity whatsoever."[3]

In the year following that retort, 1919, center and right-wing parties called for the expulsion of East European Jews, and the Prussian Interior Minister Wolfgang Heine aired the possibility of sending the "undesirables" to camps outside the cities. The German National People's Party called for a halt to the immigration of Jews, who were allegedly arriving "daily by the hundreds and the thousands [. . .] today the Jews feel themselves to be rulers of

Germany."[4] This refugee crisis—the first but not the last Einstein would witness—was caused by the dislocations of the war and the shifting of borders in eastern Europe, the dissolution of the Austro-Hungarian Empire, and the Bolshevik Revolution in Russia. To Einstein, the Jews were the victims, and he defended them in an article entitled "Immigration from the East." In it he openly supported a form of cultural Zionism: establishing a home for students and academics in Palestine. Einstein's commitment to the less privileged and the displaced would animate his actions for the rest of his life. He continued to engage in efforts on behalf of Jewish emigrants after the U.S. Immigration Act of 1924 practically erased the chances that Eastern European Jews could emigrate there. (The American situation did not improve under the administrations of Franklin D. Roosevelt; 400,000 allowed-entry permits for European refugees remained unfilled even before World War II made emigration impossible.) Einstein even became involved in plans to resettle Jews in Peru and elsewhere.

Establishing a Hebrew University in the British Mandate of Palestine, which opened on April 1, 1925, was one of Einstein's most consistent commitments. He had laid the cornerstone to the university during his sole visit to Palestine when returning by ship from Japan in 1923. He now hoped it might become "a haven for academic freedom and tolerance," as he wrote to Chaim Weizmann, head of the Zionist Organisation. "Let it never be closed to any serious intellectual striving; let no one be forbidden entry. May our tradition ever stimulate us, never hamper us."[5] His optimism of the early 1920s regarding Jewish blossoming in Mandate Palestine was severely tested in the summer of 1929, when in the aftermath of eight days of violent demonstrations, riots, and murders in Hebron, Safed, and Jerusalem, 133 Jews and 116 Arabs were killed, with many more injured. The dead Jews were victims of Arab attacks on their homes. The British colonial forces, together with some members of the Jewish Haganah (a Zionist paramilitary organization), were responsible for the Arab deaths.

Einstein attended the cornerstone laying ceremony of the Hebrew
University in Jerusalem, February 1923. *Courtesy of the Albert Einstein
Archives, The Hebrew University of Jerusalem.*

Einstein made a public statement two days after the terrible events. Work toward peaceful coexistence and mutual trust, he noted, had to carry on. He criticized what he saw as the inaction of British forces in protecting the safety of Jews. In this and subsequent statements and letters, Einstein began to carve out an independent position. He agreed with much that the Zionists proposed, but he warned that without honest cooperation and negotiations "we [Jews] will have learned nothing from our two-thousand year ordeal and deserve the fate that will befall us."[6] He engaged in correspondence with Azmi al-Nashashibi, the Arab editor of the English-language newspaper *Falastin*, and published two letters that reiterated the need for cooperation, called for the Arab population to recognize the right of Jews to live in Palestine, and made suggestions for joint administrative, economic, and social organizations. This proposal was opposed by Weizmann as undermining the Zionist leadership's platform. Einstein's sympathetic support for Brit Shalom, the Zionist intellectual movement that Hugo Bergmann co-founded, dedicated to establishing a bi-national state in which Arabs and Jews would have equal rights, was also poorly received in a Palestine reeling from the violence. The British government curtailed Jewish immigration to Palestine, banned land purchases, and imposed stricter military control.

In March 1930, Einstein wrote to his sister Maja that "our Jews have revealed themselves as chauvinistic nationalists without psychological instinct and sense for equity in the matter of Palestine-Arabs. Good thing that they are powerless and don't have cannons."[7] Einstein lost confidence that peaceful solutions would ever be achieved, although he continued to repeat his entreaties and admonitions. These concerns strengthened his commitment to fighting militarism in general. Throughout he remained clear-sighted about what he could accomplish: "One cannot chop much wood with a razor," he mused to his son about his own outspoken pacifism.[8] Militarism and chauvinistic nationalism would henceforth be entwined for Einstein—although he would never waver in his view of the justness of the enterprise of building up the Jewish community in Palestine.

The West

Already in 1921, Einstein considered leaving Berlin for good. He even contemplated abandoning academia and joining the engineering firm of his friend H. Anschutz-Kaempfe. He came to believe that the Berliners wished to hold on to him because of his "brain juice" and also because he had become a sort of "fetish" to Germans, playing the role of the holy bones of a saint to be preserved in a church.[9] As he wrote to his children in Switzerland a decade later:

> Things are looking ominous here, and I think we're getting into a kind of Mussolini craze which will probably not be long in coming. Bankruptcy and distress are dreadful, and the feeble leaders slip from one embarrassed action into another. And in addition, they don't have the strength to oppose the boom-boom one-two-three-patriots who would already have been in power if they weren't feeling better in the opposition. . . . A violent overthrow seems unstoppable. Of course, my pacifist extra-curricular activities create many enemies for me here, where you don't want to give up flirting with fishing in troubled waters. I feel I am obliged, however, to use my great name for such a necessary cause. . . . I must seek to transfer the center of my earthly existence on a less volcanic soil, i.e., to move abroad before it is too late, and I am glad that you are all sitting in Switzerland, where everything that may come will play out in a milder form than here.[10]

Einstein was sadly proven right, as emigration was the only sure route to survival for Jews in Central and Eastern Europe.

Einstein openly declared himself an "unconditional antifascist" in 1928, when the Nazi Party had won some 3 percent of the

Reichstag seats.[11] Two years later, deep resentments against the Versailles Treaty, the devastating financial crash of 1929, and crushing unemployment led to a stunning electoral showing of 18 percent of votes for Hitler in September 1930, who had campaigned on a platform of full employment. Even though Einstein was well aware of the Nazi racial theories, in public he minimized their danger to German Jews. He placed Hitler's victory in the context of the economic crisis and urged his fellow Jews not to despair: "I hope that as soon as the situation improves, the German people will also find their road to clarity."[12]

Privately, his thoughts turned ever westward, seeing Europe as increasingly a lost cause. In 1931, on board the *SS Portland* headed from Portsmouth, England, to Long Beach harbor for his second visit to Caltech, Einstein wrote in his diary:

> 6 December. We left the Channel already yesterday. It is getting continuously warmer in rainy weather and a dramatic sea. Life is enviably contemplative. Began reading Friedell's clever Cultural History volume III and Grünberg's tales [on South America]. In addition, am also reading around in Bohr's quantum mechanics. Today I decided to essentially give up my position in Berlin. Therefore, a migratory bird for the rest of my life! Seagulls are still accompanying the ship, continuously in flight. They will join in the trip until the Azores. These are my new colleagues, but God knows they are more diligent than I and know geography better, too. How dependent is man on external things compared with such a creature![13]

The steps he took over the next year ensured that he would have a guaranteed existence and a refuge abroad.

Einstein spent the winter term of 1933 at Caltech. As usual, he was in very high demand and had a full social and professional calendar, amply illustrated and documented in the daily press. On January 30, Hitler came to power in Germany. Before leaving

Pasadena for the East Coast, Einstein declared for the press that "As long as I have any choice in the matter, I shall live only in a country where civil liberty, tolerance, and equality of all citizens before the law prevail."[14]

Thus began Einstein's exile. He spent the summer in the Belgian seaside town of De Haan (Le Coq-sur-Mer) and then traveled to England. He received numerous offers there as well as from France and Spain (the last not yet under Fascist control). He intended to return to Caltech for another visit. He left Europe for Princeton, as planned, on October 7, 1933.

Einstein spent the next decade assisting refugees and providing numerous affidavits and letters of recommendation for friends, colleagues, and family members seeking to escape Europe, especially following the promulgation of the Nuremberg Racial Laws of 1935. After Germany's annexation of Austria in March and *Kristallnacht* (Night of Broken Glass) on November 9, 1938, Jews were physically in great danger of internment and murder. The Eastern European Jews—from Poland, the Soviet Union, and elsewhere—who had raised Einstein's Jewish consciousness in 1919 would bear the full brunt of Nazi brutality with the outbreak of war in September 1939.

When Einstein left Europe in the fall of 1933, his expectation was that he would not return to Germany while the Nazi Party was in power. He had expressed strong opposition to Germany's actions during World War I, and yet his commitment to scientific internationalism had made him the most influential opponent of the boycott against German and Austrian science that followed the armistice of 1918. Science was a model for how to transcend the enmities of war and build peace. As the years of Hitler's Thousand-Year Reich began, the unending litany of antisemitic measures, persecution of the disabled and defenseless, and outrages against conscience hardened him not just to the state but to the Germans in general who tolerated such a travesty.

Anger and shock followed the open violence of *Kristallnacht* in 1938, which would pale before the utter brutality of the Final

Solution: the deportation and mass execution of millions of Jews and others by the power of the state, the actions of the military, and with the acquiescence of the population. As he expressed it in a letter in 1948 to German chemist Otto Hahn, the discoverer of uranium fission and one of the few leading scientists who remained in Germany while maintaining anti-Nazi views:

> The crimes of the Germans are really the most disgusting that the history of the so-called civilized nations has to display. The attitude of the German intellectuals—considered as a class—was not better than that of the rabble. Remorse and an honest will, the least that could be done in order to redeem things that might be redeemed after the enormous murder [of the Holocaust], have not shown themselves even once. Under these circumstances I feel an irresistible aversion against being associated with any single affair that embodies a piece of German public life, simply out of a need to keep clean.[15]

Back in Princeton, Einstein remained more comfortable speaking German than English, but he reserved his native language for exiles, Russians, Jews, and Americans who had studied abroad. He never returned to the land of his birth. None of the houses where Einstein lived or studied in Germany are standing today.

The Most Famous American Citizen

Just across Constitution Avenue from the Vietnam Veterans Memorial on the National Mall in Washington, DC, is a gigantic bronze figure of Einstein: 12 feet tall and weighing 4 tons. Sculpted by Robert Berks and unveiled in 1979, the physicist lies recumbent on a granite platform, looking at a manuscript that displays the field equations of general relativity, his formula for the photoelectric effect, and of course $E = mc^2$. Although there are replicas of this monument at the Israel Academy of Sciences and Humanities in Jerusalem and at Georgia Tech in Atlanta, what stands out here is the location. Abutting the National Academy of Sciences, this is the only sculpture of a scientist on the Mall, and one of very few sculptures of identifiable individuals, joining Abraham Lincoln, Martin Luther King Jr., and Franklin D. Roosevelt. The image makes a specific statement about Einstein's belonging: he was an American.

Despite occasional grumbling about aspects of American culture and manners, Einstein was clearly impressed with the country and the potential it offered as a haven to do physics, even to an outspoken Jew with heterodox political leanings. In October 1930, in a wide-ranging article entitled "What I Believe," he commented that "my political ideal is democracy" and that the failure of democracies in Russia and Italy (it was too soon for Germany's tragedy) did not mean that the political form was doomed. "I believe that you in the United States have hit upon the right idea," he continued. "You choose a president for a reasonable length of time and give him enough power to acquit himself properly of his responsibilities."[16]

Nonetheless, Einstein had criticisms of the American political system, chiefly the treatment of African Americans. The physicist's first public comment on this issue came in response to a direct appeal by W. E. B. Du Bois, founder of *The Crisis*—the magazine of the National Association for the Advancement of Colored People (NAACP)—asking for a statement on the periodical's twenty-first birthday. Einstein, then in between his visits to Caltech and actively contemplating a move across the Atlantic, began his response by noting: "It seems to be a universal fact that minorities, especially when their individuals are recognizable because of physical differences, are treated by the majorities among whom they live as an inferior class."[17]

This thought would become a hallmark of Einstein's many statements on behalf of civil rights: anti-Black racism was caused by white people. Einstein stated this idea most forcefully to white readers in a January 1946 article, "The Negro Question":

I am firmly convinced that whoever believes [that Negroes are inferior] suffers from a fatal misconception. Your ancestors dragged these black people from their homes by force; and in the white man's quest for wealth and an easy life they have been ruthlessly suppressed and exploited, degraded into slavery. The modern prejudice against Negroes is the result of the desire to maintain this unworthy condition.[18]

This position is an adaptation of views he developed about anti-semitism back in Germany. Antisemitism likewise was not the fault of the Jews but rather of those who profited from the imposition of difference. His self-identification as a member of a marginalized people served as a template for him to advocate for others whom he saw as analogously situated.

Einstein's location in Princeton accentuated his deep commitment to civil rights for African Americans. That small town was decisively segregated, although without the explicit codification prevalent in the Jim Crow South. Einstein took many opportunities to cross into Black neighborhoods to spend time with residents on their porches. More publicly, he antagonized local elites in 1937 by inviting the great contralto Marian Anderson, who had just performed at the local McCarter Theatre, to stay at his home when she was refused a room at the whites-only Nassau Inn. Einstein had similarly walked backstage to meet singer and actor Paul Robeson, a Princeton native, at the McCarter in 1935, beginning an extended relationship bonded as much by politics as by music. Einstein signed on to Robeson's American Crusade to End Lynching and continued to support his causes despite the consternation of others that stemmed not least from Robeson's open affiliation with the Communist Party. Resistant in his later years to accepting honorary degrees, in 1946 Einstein gladly accepted one from Lincoln University in Pennsylvania, the first college-degree granting historically Black university in the United States.

Einstein's statements on civil rights were part of the reason why J. Edgar Hoover, director of the Federal Bureau of Investigation, opened what grew into a gargantuan file on Princeton's most famous resident. The physicist made a promising target. In 1949, when any left-wing sympathies could generate unwelcome scrutiny, Einstein published in the first issue of the American Marxist periodical *The Monthly Review* an article entitled "Why Socialism?" "The economic anarchy of capitalist society as it exists today is, in my opinion, the real source of evil," Einstein wrote. "I am convinced there is only *one* way to eliminate these grave evils, namely through the establishment of a socialist economy accompanied by an educational system which would be oriented toward social goals."[19] Hoover's file expanded at a rapid clip. Full of innuendo and hearsay, fueled by wiretaps and the opening of Einstein's mail, the file demonstrates an obsession with labeling the creator of general relativity a communist. In 1950, Hoover tried, unsuccessfully, to have Einstein's American citizenship revoked.

Einstein's adherence to some form of socialism went back many years, but so did his suspicion of Soviet communism, a subtlety that eluded his right-wing detractors. Einstein declined many invitations to visit the Soviet Union. He had an abiding suspicion of the Bolshevik regime due to both his presumptions about Russian antisemitism and the increasingly violent despotism that emerged under Joseph Stalin in the 1930s. In 1946, four Soviet scientists wrote a public letter to Einstein castigating his support for world government as a solution to the nuclear arms race while acquiescing in the United States' withholding of nuclear information from the Soviet Union. In his response, Einstein advocated turning nuclear secrets to the United Nations instead of any state.

Einstein was to some extent insulated from the worst excesses of the persecutions and Red-baiting known as McCarthyism, which reached full flower in the early 1950s. He did not repudiate his former left-wing associations, and he advised those subpoenaed by the House Un-American Activities Committee, such as the high school teacher William Frauenglass, to refuse to testify not on the grounds of the Fifth Amendment (protection against self-incrimination) but on those of the First (freedom of speech). He advocated civil disobedience in the face of anticommunist hysteria. He was never summoned before any of the prosecuting committees, and so he never had to demonstrate his willingness to defy them. His 1953 public defense of Frauenglass and his opposition to the execution of Julius and Ethel Rosenberg for atomic espionage elicited a deluge of new correspondents that year; Einstein chose not to answer many of these letters, which included a significant amount of hate mail.

Consider the contrast to J. Robert Oppenheimer, the director of wartime Los Alamos, the project that had designed America's first atomic bombs, and after the war Einstein's boss as director of the Institute of Advanced Study. During the 1930s, as a professor of theoretical physics at Berkeley, Oppenheimer openly associated with communist causes and individuals, including his brother Frank and several of his graduate students who were Party members. During and after the war, Oppenheimer was at pains to distance himself from these ties, especially as he headed the advisory apparatus of the Atomic Energy Commission. This was not enough for some hardliners, however, who in 1954 stripped Oppenheimer of his security clearance.

Despite sometimes tense relations with Oppenheimer, who considered Einstein a has-been who had lost his acumen over quantum mechanics, the sage of Princeton was openly sympathetic to him after his public disgrace. As Einstein said in a phone conversation with his close friend Hanna Fantova on April 13, 1954: "The Atomic Energy Commission is all in a tizzy because McCarthy has attacked it. Oppenheimer is a black sheep and I do not understand why he has not long ago abandoned the entire affair. He was only an advisor. In his position I would have said as the Saxon King did: 'Deal with your garbage yourselves!'" A few weeks later, he told Fantova that he "was over at Oppenheimer's to console him; he takes it very tragically and is swimming—he sees only water everywhere. I do not at all understand why Oppenheimer takes the situation so seriously."[20] Einstein was disgusted by the hysteria. As he wrote to a journalist in 1954: "I would rather choose to be a plumber or a peddler in the hope to find that modest degree of independence still available under present circumstances."[21]

Einstein had recently had an opportunity to demonstrate his sincerity in forswearing conventional politics. On November 9, 1952—the fourteenth anniversary of the tragedy of *Kristallnacht*—Chaim Weizmann passed away. A biochemist by training, he died as the first president of the State of Israel, which was born in 1948 of both United Nations resolutions and armed conflict. Political power in Israel resided with David Ben-Gurion, the prime minister, but the president was the official head of state. Ben-Gurion's personal secretary, Yitzhak Navon, recalled the prime minister saying immediately upon receiving the news of Weizmann's death: "There is only one man whom we should ask to become the President of the State of Israel. He is the greatest Jew on earth. Maybe the greatest human being on earth. Einstein."

One week later Ben-Gurion sent a cable to Abba Eban, Israel's ambassador to the United States, asking him to contact the physicist. Privately, Ben-Gurion was of two minds. After he dispatched the cable, he told Navon: "Tell me what to do if he says yes! I've had to offer the post to him because it's impossible not to. But if he accepts, we are in for trouble." Einstein, divining the topic, returned Eban's call and forestalled him by saying that he would not take a position for which he felt himself unsuited. Eban persuaded him to at least think it over and went to Princeton to meet with Einstein. Again, he declined. Eban recalled that he had "the impression that the offer has not caused him any elation or pride, but rather sorrow."[22] Einstein felt strongly about Israel, but he did not belong there. He never visited the new country.

Chapter 5

War and Peace

In 1901, Albert Einstein presented himself to a draft board. He had already renounced his German citizenship in part to avoid serving in the military, but he was now a citizen of the Swiss Canton of Zurich, and military service was an obligation. His aversion to militarism acquired in adolescence had not abated. In 1912 he wrote to a friend of his Serbian wife, Mileva Marić, lamenting Habsburg "saber rattling" portending a potential "conflict with Austria [which] would be bad for the Serbs, even in victory."[1] It is therefore somewhat hard to picture what would have happened had he been enrolled in the Swiss army, but we do not have to imagine. He was deemed unfit for service due to "*Varices*" (varicose veins), "*Pes Planus*" (flat feet), and "*Hyperhidrosis ped.*" (excessive foot perspiration).[2] He did not denounce war as a barbarism while living in Zurich.

In April 1914, Einstein moved from Zurich to Berlin, taking up his post at the Prussian Academy of Sciences in Berlin, the pinnacle of German physics. His dominant concern, however, was the disintegration of his first marriage. After a summer of deep unhappiness and strife, Mileva Marić and their two sons returned to Zurich on July 29, 1914. A month earlier, Habsburg Archduke Franz Ferdinand had been assassinated by a Serbian nationalist, and on August 3 Germany joined Austria-Hungary in declaring war on Russia and France. Britain joined the defense of threatened Belgium the next day. Europe was plunged into war. The ensuing four years of savagery brought Einstein's hitherto inchoate feelings to the fore. By the time of the Armistice on November 11, 1918, Einstein had become a declared pacifist, and that pacifism would serve as a unifying bond among his various political commitments of his Berlin years and beyond.

Being a Jew and being a scientist both posed challenges to Einstein's pacifism. When Hitler's Nazi Party seized power in Germany in 1933, beginning Europe's and the globe's renewed descent into cataclysm, Einstein's first response was to affirm his pacifism. However, the scale of Nazi brutality, even before the onset of World War II, would come to shake Einstein's conviction that military violence was never a solution. At the end of that war, the advent of nuclear weapons, which the public came to identify with Einstein, both revived and significantly deepened his interwar opposition to the militarism of national sovereignty and the necessity to avoid armed conflict. Despite his continually worsening health—indeed, until his final days—he assiduously undertook the task of ridding the world of war. No draft board could excuse him from this duty.

The Great War

Only a few days after the outbreak of World War I, Einstein wrote to Paul Ehrenfest in neutral Holland: "Europe in its madness has now embarked on something incredibly preposterous. At such times one sees to what deplorable breed of animalistic brutes we belong."[3] This revulsion struck Einstein close to home. Upon arriving in Berlin, he stayed with his friend Fritz Haber, who had been instrumental in arranging his new appointment. Haber, crucial in mitigating the worst spousal conflicts between the Einsteins, was to become a dark foil to Einstein's pacifism. A much-respected physical chemist, a fervent patriot, and a Jew who had converted to Protestantism in his youth, Haber proved instrumental to Germany's war effort by organizing and implementing the introduction of chemical warfare. Haber lost much for his pains: his wife committed suicide with his military pistol in 1915, and the war consumed his close friendship with Einstein.

Einstein's aversion to militarism developed alongside the intensifying carnage. In mid-October 1914, the Berlin physician Georg F. Nicolai, in collaboration with the astronomer and co-founder of the German Society for Ethical Culture Wilhelm J. Förster, drew up a statement in response to the notorious appeal "To the Civilized World," in which ninety-three German intellectuals, artists, and academics rejected German culpability for the outbreak of the war and defended military atrocities in Belgium and elsewhere. Among the signatories of the infamous appeal were many distinguished scientists and close colleagues of Einstein, including Max Planck, Walther Nernst, and Fritz Haber. Einstein later tried to excuse their actions, emphasizing that some of the signatories had not read the document before agreeing to sign. Nonetheless, the manifesto marked a watershed for Einstein, as it did for foreign scientists, who would cite it to justify the victorious powers' years-long boycott of German and Austrian scientists after the war.

Einstein was exempted from that boycott, largely because of his involvement with Nicolai's counter-manifesto, later known as the "Manifesto to the Europeans." This was not a pacifist document; that is, it did not reject war as illegitimate. Instead, it urged that the community of European intellectuals cooperate constructively after the end of hostilities. Although the statement circulated among professors, only Einstein and Otto Buek, a philosopher and translator, were prepared to sign. The counter-manifesto is important not for its effects (it had none, being kept secret until 1917, similarly to other antiwar documents) but because it showed Einstein's willingness to embrace this very unpopular position. Incidentally, the first time the word "pacifism" shows up in Einstein's writings was in January 1920 in connection with this document, when he characterized Nicolai as a "pacifist writer."[4]

The Central Organization for a Durable Peace, newly established in the Hague to prepare for reforms in postwar international jurisprudence, welcomed Einstein's election to the German national committee. In response to one of their questionnaires, Einstein wrote in 1917 that a League of Nations then being proposed should not only devote itself to political and juridical matters but should also extend its sphere to "establishing minimum wages and other protective measures for the economically weak."[5] In 1915 he had also become a member of the pacifist organization Bund Neues Vaterland. In both signing the counter-manifesto and joining the Bund, Einstein was likely primarily concerned with the importance of international collaboration among intellectuals and scientists rather than with the abolition of war.

His closer public engagement with political causes came at the end of the war, during the workers' and soldiers' revolution in Germany. Einstein's lecture notes laconically describe that his class was canceled "due to revolution."[6] Notably on November 9, 1918, the imperial government fell amidst massive demonstrations in Berlin. That day, Einstein journeyed through chaotic Berlin with physicist Max Born and psychologist Max Wertheimer, both at the University of Berlin, to appeal for the release of professorial colleagues from the University of Berlin who had been seized by revolutionary students. In spite of his initial optimism for the newly declared German Republic, over the next years the grave deterioration into hunger, disease, and armed conflicts in the streets only impelled him to further action.

Interwar and Antiwar

In the interwar period, Einstein incrementally developed a mature, and increasingly militant, theory of pacifism. Due to his rising visibility after the successful 1919 British eclipse expedition that confirmed the predictions of his theory of gravitation, Einstein was drawn into numerous actions and appeals across Europe, often alongside intellectuals with humanistic and liberal outlooks. Yet among them, reconciliation among scientists and intellectuals emerged as his main goal. He therefore championed the newly formed League of Nations and genuinely believed in its potential to bring about a stable, peaceful coexistence, and tolerance among former enemies. By the end of the 1920s, he wrote to the pacifist poet Kurt Hiller: "My political views have not changed [recently]; earlier I did not engage publicly with political matters. This came about through inquiries and requests to which I had to react."[7]

This initial post-war, post-imperial optimism was soon tempered by the fragility of the new republic, by extremism on the right and left, by the never-abating militarism, and by the emergence of paramilitary organizations. Einstein tried to stem a hate campaign against Nicolai, who had drawn up a proclamation protesting the murder in January 1919 of socialist leaders Karl Liebknecht and Rosa Luxemburg, an appeal that Einstein also signed. The coalition of left-leaning political parties was defeated already in the elections of June 6, 1920. The political atmosphere became increasingly polarized. Among revanchist circles, being a pacifist was tantamount to treason. Prominent among the more virulent events was the murder of Hans Paasche, a former naval officer turned pacifist, in May 1920. The authors of the "Manifesto to the Europeans," Förster and Nicolai, were both fired from their positions. The mathematician Emil J. Gumbel, a pacifist, was beaten and threatened with assassination. So was Nicolai. The journalist and politician Hellmut von Gerlach, founder of the German Democratic Party, feared for his life. Colleagues informed Einstein of nationalist and reactionary outbursts at other universities such as those of Rostock and Tübingen.

The first printed call for Einstein's murder appeared in early 1921, shortly after he was accused of treason alongside other members of the Bund Neues Vaterland. Undeterred, Einstein continued to push for international cooperation and rapprochement. His trip to Paris at the end of March 1922 represented an important step, yet he agreed to undertake it only after significant urging by French colleagues. He ended up delighted with his reception in Paris. Most of the French press praised Einstein as a messenger of peace. They quoted a parliamentarian saying that "Einstein in Paris is the beginning of the healing of all international madness."[8] French physicist Paul Langevin stayed with the Einsteins on a visit to Berlin to speak at the rally of the *No More War!* movement in July 1923, but Berlin's police chief forbade him to address the public. Einstein deemed Langevin "too good for this world."[9]

Einstein invited the physicist Paul Langevin to visit Berlin, during a time when Germany was still boycotted by the former Entente powers, including Langevin's homeland, France. They attended this antiwar demonstration in Berlin in July 1923. *Courtesy The Granger Collection.*

Einstein joined the International Committee on Intellectual Cooperation
of the League of Nations—although he was a citizen of Germany, which
was not yet permitted to join—at the insistence of French scientist Marie
Curie. This image shows Einstein and Curie with fellow Nobel Laureate
Robert A. Millikan, the American representative to the committee,
in Geneva in July 1924. *Courtesy of the Archives, California Institute of
Technology.*

Einstein had become a member of the International Commission on Intellectual Cooperation of the League of Nations at its
inception in January 1922, even though Germany would not be
permitted to join the League for four more years. He made that
decision at substantial personal risk. Einstein's friend, German
Foreign Minister Walther Rathenau—who had encouraged him to
travel to Paris—was assassinated by a right-wing paramilitary in
June 1922, devastating Einstein.

By the end of this decade, Einstein's activism for peace extended
beyond the concerns of scientists, such as international cooperation, or of German citizens, such as the questions of war guilt, reparations, and the normalization of relations. His most controversial

intervention on the topic of war resistance was his "Two-Percent" speech on "Militant Pacifism," delivered in the United States on December 14, 1930, to the New History Society, a dissident faction of the Baha'i movement. Speaking in German, Einstein said: "If you can get only two percent of the population of the world to assert in time of peace they will not fight" there would be "not enough jails in the world to accommodate them!" He explained that he was raising funds for organizations to support those who resisted conscription "not prompted by selfish or cowardly motives," but because they were convinced that war was wrong.[10] Their attitude might put a stop to war.

Einstein had arrived in New York harbor, on his way to Caltech, aboard the ocean liner *Belgenland*, three days before the speech and was greeted by close to one hundred reporters and photographers. He made several addresses that morning, among them a live "Greeting to America" over the National Broadcasting Corporation radio network, stating that: "It is in your country, my friends, that those latent forces which eventually will kill any serious monster of professional militarism will be able to make themselves more clearly and definitely felt. Your political and economic condition today is such that if ever you set your hand to this job in all seriousness you can entirely destroy the dreadful tradition of military violence under which the sad memories of the past, and to a certain extent the whole of the world, continues to suffer even after the last great warning of the Great War."[11]

These addresses, delivered at the start of only his second visit to the United States in a decade, did not endear Einstein to everyone who listened on the radio and confronted the deluge of newspaper coverage. The aftershocks of his outspokenness on this matter would reverberate over the coming years. In 1932, the Women's League of America denounced him as a dangerous communist apt to damage national security, thereby jeopardizing his entry visa for his third visit to Caltech the following year. The FBI took note. Even in increasingly isolationist America, pacifism was threatening.

World War II and Nuclear Weapons

When the city of Hiroshima was devastated by the first atomic bomb at 8:16 in the morning on August 6, 1945, Albert Einstein did not know. Nor was he alerted when a sustained fission chain reaction was initiated in the reactor underneath the football field at the University of Chicago in 1942. Nor was he advised of the first nuclear explosion on July 16, 1945. He learned of the creation and use of atomic bombs by reading the newspapers and listening to the radio shortly after the destruction of Hiroshima. As soon as Einstein became aware of the successful development of atomic bombs, he mobilized his scientist colleagues and political contacts to lobby for international control of nuclear weapons. Yet within a year Einstein was so indelibly associated with nuclear weapons and the destruction of two Japanese cities that today it seems impossible to disentangle them. The connection took months to develop, and Einstein resisted until resistance became futile. That point arrived before July 1, 1946, when Einstein appeared on the cover of *TIME* Magazine, his iconic mustache and hair juxtaposed with the mushroom cloud over Nagasaki, its crown featuring a ghostly $E = mc^2$. The caption: "COSMOCLAST EINSTEIN: All matter is speed and flame."

Given Einstein's record of public opposition to military conflict, it surprised many to learn in August 1945 about the letter he had sent in August 1939 from his summer residence at Peconic, Long Island, to President Franklin D. Roosevelt. The two-page note informed the president that recent research by Frédéric Joliot-Curie in France and Enrico Fermi and Leo Szilard in the United States—both of the latter being refugees from Mussolini and Hitler, respectively—suggested "that it may become possible to set up a nuclear chain reaction in a large mass of uranium" and thereby create a potentially vast power source. The phenomenon of the fission of uranium, discovered only the previous December in Berlin, could "also lead to the construction of bombs, and it is possible—though much less certain—that extremely power-ful bombs of a new type might be constructed. A single bomb of this type, carried by boat and exploded in a port, might very well destroy the whole port together with some of the surrounding ter-ritory."[12] Einstein warned Roosevelt that German scientists led the way in this domain and that the Nazis had recently seized control of the rich uranium deposits in Czechoslovakia.

This is Einstein's most famous letter. Szilard and Fermi, together with their fellow émigré, Princeton physicist Eugene Wigner, encouraged him to alert Roosevelt. Einstein dictated a first draft to Wigner over the telephone. After revisions and consultations that also included Hungarian émigré Edward Teller, a typed version was brought to Einstein for his signature. It was then passed on to the financier Alexander Sachs, who promised to hand it to Roosevelt. Events in Europe galloped ahead. Germany invaded Poland, the "phoney war" began, and Sachs did not have the opportunity to meet Roosevelt until mid-October. The president read the letter immediately and organized a committee to discuss next steps. This was autumn of 1939, less than a year after the discovery of uranium's unexpected properties.

Einstein's letter did trigger research into fission-based weap-ons over the next two years, but it was not yet the frenetic activity that one associates with the Manhattan Project. Roosevelt first impaneled a "Uranium Committee" under the leadership of the Director of the Bureau of Standards, the engineer and physicist Lyman J. Briggs. With little urgency and even fewer resources, the group's accomplishments were meager. Einstein was, however, kept apprised of developments from time to time.

Across the Atlantic, the pressure was significantly higher. The British and their complement of scientific émigrés from Hitler's Reich explored the practicalities of building a nuclear explosive in a more determined fashion. In March 1940, Otto Frisch and Rudolf Peierls calculated a "critical mass" for the creation of a runaway chain reaction of a specific isotope of uranium (U-235) and determined that a bomb would be feasible. This highly classified report was transmitted to Roosevelt's scientific advisors, Vannevar Bush and James Bryant Conant, who then lobbied for a radical acceleration of research into nuclear weapons. Joseph Stalin also received a copy of the Frisch-Peierls report through the British intelligence officer and spy John Cairncross.

Bush, the chairman of the National Defense Research Committee, and committee member Conant, the president of Harvard University, had their way. Even before the Japanese attack on Pearl Harbor of December 7, 1941, the Americans were poised to reassign fission research to the military. The official transfer to the U.S. Army Corps of Engineers, Manhattan District—the highly secret "Manhattan Project"—took place on August 13, 1942, more than three years after Einstein's letter. It should surprise no one that the orator of the "2% speech" was not granted security clearance to work on the Manhattan Project. Three years after the project was created, two Japanese cities would be destroyed by its work.

Einstein was among the first scientists contacted by journalists after the news from Hiroshima broke on August 6, 1945. At his summer cabin in upstate New York, Einstein was quite reticent to discuss the shocking new weapon, demurring from commenting "at present" to the *New York Times* on August 8.[13] The next day, news arrived of a second bomb destroying the city of Nagasaki. Two days later, Einstein gave a thirty-minute interview in which he explained that "atomic power is no more unnatural than when I sail my boat on Saranac Lake." He stated: "I have done no work on the subject, no work at all. I am interested in the bomb the same way as any other person; perhaps a little more interest. However, I do not feel it justified to say anything about it."[14]

The following day, on August 12, 1945—halfway between the Nagasaki attack and Japan's unconditional surrender—the public unveiling of the science behind the American development of the atomic bomb placed Einstein front and center. The U.S. government released *Atomic Energy for Military Purposes*, an official account of the Manhattan Project composed by Henry DeWolf Smyth, chair of the physics department at Princeton University. Carefully curated to exclude all classified material, the report highlighted nuclear physics over chemistry and engineering. It also highlighted Einstein. In the fourth paragraph of the book, we encounter the first equation: $E = mc^2$. "Even Einstein could hardly have foreseen its present applications," Smyth wrote, "but as early as 1905 he did clearly state that mass and energy were equivalent."[15] Smyth's account of atomic bombs was reprinted, translated, and excerpted around the globe.

On August 27, 1945, three weeks after the bombing of Hiroshima, Einstein, still at Saranac Lake, wrote to the celebrated ABC broadcaster Raymond Swing—who devoted one newscast a week to the dangers of the atomic bomb—regarding his radio broadcast four days earlier. Einstein expressed his "deep gratefulness for, and agreement with" Swing's "systematic endeavor to enlighten public opinion in favor of the creation of effective world-government. You have rightly pointed out that the occupation of the Pacific Islands by the U.S.A. alone is a step in the wrong direction which must have unfavorable consequences for all future attempts to bring about international security. But the worst of all is the upholding of military secrecy and the maintenance of huge organizations to build up new secret weapons on a national scale.... I am astonished really, that this public is not reacting in a stronger way seeing the great dangers and the fateful mistakes of our government."[16]

Swing, pleased by the letter, met Einstein in Princeton in September. A transcript of remarks from that conversation was published in November 1945 in the *Atlantic Monthly* as "Einstein on the Atomic Bomb. Albert Einstein as Told to Raymond Swing." The editor's prefatory note quoted Einstein's letter to Roosevelt and declared that it was Einstein's "daring formula" which led to the concept that "atomic energy would some day be unlocked."[17]

The *Atlantic* piece was Einstein's first direct public discussion of the atomic bomb. His subsequent statements until his death in 1955 largely continued its themes, which in turn drew upon his longstanding views on war. "The release of atomic energy has not created a new problem," the article began. "It has merely made more urgent the necessity of solving an existing one." Although he commented on some of the topical issues of the moment—the role of the fledgling United Nations and the question of whether the Soviet Union should be given "the secret of the bomb"—he continued to maintain his hope for a world government that would constrain military sovereignty. He also reiterated his distance from the Manhattan Project: "I do not consider myself to be the father of the release of atomic energy. My part in it was quite indirect. I did not, in fact, foresee that it would be released in my time."[18]

Yet it had. Most of Einstein's actions following the atomic bombings were not those of a physicist but of a pacifist. He became involved in the atomic scientists' movement, a loose affiliation of Manhattan Project veterans, political theorists, and other concerned individuals seeking to navigate in a world they saw as unalterably changed by the apocalyptic potential of nuclear weapons. This constellation soon fragmented into separate groups tackling distinct aspects of the problem.

In May 1946, Einstein issued an appeal for contributions to a $200,000 fund for a nationwide campaign "to let the people know that a new type of thinking is essential" in the atomic age "if mankind is to survive and move toward higher levels." The campaign was to be carried out by the newly formed Emergency Committee of Atomic Scientists—which included prominent scientists such as Hans Bethe, Leo Szilard, and Edward Condon—chaired by Einstein. (The group was officially incorporated in New Jersey on the one-year anniversary of the destruction of Hiroshima.) In this appeal Einstein, for the first time, included himself among those responsible for the Atomic Age: "We scientists who released this immense power have an overwhelming responsibility in this world life-and-death struggle to harness the atom for the benefit of mankind and not for humanity's destruction."[19] Einstein's numerous appeals and soliciting of financial support for the Committee's actions drove him to acquire an auto-pen to affix his signature to the hundreds of outgoing letters and responses.

Even so, Einstein was slow to condemn the bombings of Hiroshima and Nagasaki. It was only on August 19, 1946, that he stated to the *New York Times* that he "deplored" the use of the bombs and that he was convinced that Roosevelt, who had died in April 1945, before the first bomb was tested, would not have dropped them.[20] Late that month, the *New Yorker* magazine devoted an entire issue to journalist John Hersey's account of six survivors of the first atomic bombing—released within two months as the now canonical book, *Hiroshima*—that changed the understanding of atomic war for a global audience. Hersey provided a searing depiction of gruesome devastation and loss, and public opinion began to question the morality of these devices, boosting the atomic scientists' movement and Einstein's Emergency Committee. The coming years would see significant efforts by the Truman administration to diminish the impact of Hersey's salvo that marked the end of the first nuclear year.

By this point, Einstein had become firmly associated with the nuclear age. On the first of July, the Americans had detonated the first nuclear explosion since Nagasaki at the Bikini Atoll. Invited international observers, pointedly including the Soviets, viewed the two atomic blasts. Some spectators were awed by the horror of it; others were underwhelmed—neither reaction was conducive to taming the plutonium genie. On June 14, the U.S. envoy to the United Nations Atomic Energy Commission, Bernard Baruch, had unveiled his eponymous plan for nuclear disarmament—on American terms, as Soviet ambassador Andrei Gromyko was quick to point out. Debates were underway as the nuclei fissioned over the Marshall Islands, but they soon were rendered, perhaps predictably, fruitless. The proposal was dead by early 1947. The Emergency Committee became increasingly rudderless after the Baruch Plan's failure, and it became inactive by 1949.

On February 12, 1950, Einstein was the first guest on the half-hour "Today with Mrs. Roosevelt" show, Eleanor Roosevelt's first television program, which aired on NBC for three months on Sundays at 4:00 p.m. In response to the crash research program to develop a hydrogen bomb (powered by thermonuclear fusion, rather than the fission of heavy elements like uranium) initiated by Truman two weeks earlier, Einstein expressed his opposition to such a Super bomb: "If successful, radioactive poisoning of the atmosphere, hence the annihilation of any life on earth, has been brought within the range of technical possibilities."[21]

The residents of Hiroshima and Nagasaki experienced the devastation of the bombs firsthand, but the rest of the Japanese public learned the details of the 1945 bombings only seven years later. The U.S. occupation of Japan ended in April 1952, lifting the official censorship that forbade any discussion of nuclear weapons. Specifics about the magnitude of the devastation and about radiation victims now began to emerge. Earlier, two Japanese students had written to Einstein asking for a guiding principle for their studies and careers in science; in 1950, the Japan Academy elected Einstein an honorary member. Now, Japanese citizens demanded that Einstein reconcile his pacifism with his involvement in atomic devastation.

The first direct challenge regarding the atomic bombing of Japan came in 1952 from *Kaizo*, the publishing house upon whose invitation Einstein had spent six weeks in Japan in December 1922. Back then, when *Kaizo*'s president asked Bertrand Russell to name the three greatest people in the world, he replied: "First Einstein, then Lenin. There is nobody else."[22] Now *Kaizo* wrote to ask Einstein about the bombings and the coming "nuclear age." His one-page response, written on September 20, 1952, was publicized widely but inaccurately. He reiterated that his only involvement with the nuclear project was to write the letter to Roosevelt in which he "stressed the necessity of large-scale experimentation to ascertain the possibility of producing a nuclear bomb." While "well aware of the dreadful danger for all mankind," Einstein claimed that "the probability that the Germans might work on that very problem with good chance of success prompted me to take that step. I did not see any other way out, although I always was a convinced pacifist."[23]

The Japanese pacifist Seiei Shinohara, who translated Einstein's German text for the journal, subsequently wrote to Einstein to point out inconsistencies in his pacifism. They wrestled over the phrase "absolute pacifist": Einstein explained that he was a "convinced pacifist" and that there were conditions under which he felt it was "required to apply force, namely, in the case of an opponent whose unconditional goal it is to destroy me and mine."[24] That position was itself a modification that Einstein had reached in the 1930s in view of the escalating violence of Hitler's Third Reich. Shinohara countered that Einstein's position was being used as an argument in favor of rearming Japan, and that Einstein's "conditional pacifism" was no pacifism at all.[25] Einstein stood his ground: the moral necessity of acting against German proliferation did not condone the bombings of Japan, just as Shinohara (who had spent the war years in the Third Reich) could not be held responsible for the actions of the Japanese military in China and Korea.

Ending in January 1955, it was also one of his last exchanges on nuclear weapons. His final statement was the "Einstein-Russell Manifesto," which he signed on April 11. The manifesto was the product of Russell's collaboration with physicist and peace activist Joseph Rotblat, the only scientist who had left the Manhattan Project on moral grounds. It was released on July 9. Einstein's signature appeared posthumously, alongside those of eleven leading intellectuals, including the Japanese physicist Hideki Yukawa, reacting to the terrifying dangers posed by the exponential advance in destructiveness posed by the recent advent of hydrogen bombs. Its central message was resonant with those he had been developing in the decade before the discovery of nuclear fission: "The abolition of war will demand distasteful limitations of national sovereignty."[26] That message, not $E = mc^2$, was the idea that Einstein wanted humanity to take forward into the Atomic Age.

Chapter 6

Free Creations

On June 10, 1933, just past the midpoint of his adult life and at a personal and scientific fork in the road, Einstein stood before a crowded audience in Rhodes House, Oxford. He had recently crossed North America and the North Atlantic as he wended his way from his visiting stint at Caltech to Belgium, whose reigning queen was a friend, and then on to the United Kingdom. Germany was no longer safe: Adolf Hitler's National Socialists had seized unconstrained power in March and had quickly transformed the country into a racist dictatorship that saw Einstein as a public enemy who needed to be eliminated. Given intelligence that he was targeted for foreign assassination, he was accompanied by bodyguards during this visit to Britain. He had already made the decision to settle in Princeton, and he took this opportunity to say farewell to the Old World before he settled permanently in the New. To mark the change, he delivered his talk in English.

Einstein was in Oxford to deliver the Herbert Spencer Lecture. Given the highly charged personal and political circumstances, his choice of topic might strike one as odd: "On the Method of Theoretical Physics." Rather than expound on Germany's tragedy, or on pacifism, or on the plight of the Jews in Europe and in Mandate

A few months after his Herbert Spencer lecture, Einstein spoke at an assembly at the Royal Albert Hall in October 1933 to raise funds for refugees from Nazi Germany. From the left, he is accompanied by Oliver S. Locker-Lampson, the member of Parliament who personally assured Einstein's safety during this final visit to Europe, nuclear physicist Sir Ernest Rutherford, and Sir Austen Chamberlain, member of Parliament. *Courtesy of the Albert Einstein Archives, The Hebrew University of Jerusalem.*

Palestine—all topics that had occupied much of his thought of late—he turned to an account of the fundamental concepts undergirding science. Stalled on his quest for a unified field theory and beset by hostile criticism from the partisans of Bohr's interpretation of quantum mechanics, he may have regarded this exposition of method to be a valediction not just to Europe but also to his astonishing stretch of scientific productivity.

At many moments in his life—some pivotal for his scientific work, others less so—Einstein had turned to the philosophy of science. But it was not always the same philosophy. Replying to some scientific critics in 1949, sixteen years after the Spencer Lecture, Einstein insisted that although a scientist was grateful for epistemological analysis, "the external conditions, which are set for him by the facts of experience, do not permit him to be too much restricted in the construction of his conceptual world by the adherence to an epistemological system." The result, he wrote, was a kind of oscillation among systems that made a scientist appear like an "unscrupulous opportunist."[1] Nonetheless, thoughts about epistemology litter his correspondence, his scientific papers, and his writings on topics ranging from pacifism to history. While we cannot therefore consider Einstein's comments in Oxford as the be-all and end-all of his scientific credo, it is important to reflect that at a moment of crisis he turned not just to an exposition of the philosophy of science, but to this particular incarnation of it.

What, Einstein asked, can we know about nature? Science consisted of the combination of two factors: empirical evidence obtained through observation and experiment, and reasoned reflections upon that material. Assessing the relative importance and precise interaction of these factors had been the animating questions in the Western tradition of the philosophy of science since Plato and Aristotle. Einstein had made his own assessment of the balance. "Reason gives the structure to the system; the data of experience and their mutual relations are to correspond exactly to the consequences in the theory. On the possibility alone of such a correspondence rest the value and the justification of the whole system, and especially of its fundamental concepts and basic laws," he noted to his Oxford audience. "But for this, these latter would simply be free inventions of the human mind which admit of no *a priori* justification either through the nature of the human mind or in any other way at all."[2]

That phrase—free inventions (or free creations) of the human mind—cropped up again and again in the various, sometimes quite different, formulations of Einstein's epistemology. He had publicly unfurled this idea for the first time in "Geometry and Experience," a lecture he delivered to the entire Prussian Academy of Sciences in 1921, to describe the axioms of geometry. Ironically, just a few months before the Spencer Lecture, the Nazis had pressured that same Berlin Academy to rescind Einstein's membership, which he had held since 1914. As it happened, he had already resigned in disgust days earlier, denying Hitler's minions the satisfaction of evicting him. Einstein's notion of free invention pointed to an essential creativity in scientific thought: ideas were put forward by individuals, which were then tested against the evidence of our senses, but they were not necessarily derivative of that evidence. Theories were an act of human assertion, of the will to understand.

Isaac Newton, Einstein told his British audience in 1933, believed that he could derive his laws of motion from experience; yet he had not in fact done so. Newton felt compelled to use the idea of absolute space to buttress his physics, and his theory was so successful that it never occurred to him or his immediate successors to explore the fictitiousness of that idea. "On the contrary, the scientists of those times were for the most part convinced that the basic concepts and laws of physics were not in a logical sense free inventions of the human mind, but rather that they were derivable by abstraction, i.e. by a logical process, from experiments and sensory experience. It was the general theory of relativity which showed in a convincing manner the incorrectness of this view."[3]

General relativity demonstrated that an unmodifiable, Euclidean space was an unnecessary hypothesis. In Oxford, Einstein presented his own theory of gravity as a consequence of mathematical reasoning, an expression of intellect that was both tested by and gave order to the chaos of sense experience. "Experience can of course guide us in our choice of serviceable mathematical concepts; it cannot possibly be the source from which they are derived; experience of course remains the sole criterion of the serviceability of a mathematical construction for physics, but the truly creative principle resides in mathematics."[4] Einstein's actual path to general relativity involved quite a bit of tacking between mathematical and physical approaches, but by the early 1920s he typically preferred to emphasize its mathematical structure almost exclusively.

We can better appreciate the radical force of Einstein's position in the Spencer Lecture by pointing to two leading philosophical traditions he was at pains to oppose. The first was that of the eighteenth-century Königsberg philosopher Immanuel Kant, who had recently been revitalized and modified by a school of "neo-Kantians." That group argued that Einstein's redefinition of the fundamental categories of space and time could still be compatible with Kant's philosophical framework—in which Euclidean space and absolute time were taken as a priori. For Kant, those needed to be given to allow us to make sense of the phenomena of the world. Einstein had exploded that assumption. Although he was not unsympathetic to the neo-Kantians' efforts, he carefully marked himself apart in a series of reviews of Kantian interpretations of his work in the mid-1920s.

Even more significant was Einstein's engagement with what would become the dominant philosophy of science of the century: logical empiricism. Einstein had multiple engagements with the founders of this school of thought, who gravitated around the charismatic Moritz Schlick, a correspondent of Einstein's for years regarding both epistemology and pleas for career help. They shared an inspiration from Austrian physicist and philosopher Ernst Mach. Mach's analyses of mechanics (1883) and thermodynamics (1896) had argued that only concepts which were directly tied to empirical data should have a place in physical theory. This viewpoint proved formative for the young Einstein's dismissal of the ether in his 1905 theory of special relativity. While debating with his friends in the Olympia Academy in Bern, the "unscrupulous opportunist" in Einstein found much that was valuable in Mach. Yet Mach, who died in 1916, would in his final years come to express doubts about general relativity and atomism, precipitating a rupture with Einstein.

In 1922, with assistance from Einstein, Schlick succeeded to Mach's chair at the University of Vienna, where he joined and soon led a local group of philosophers and scientists who came to be known as the Vienna Circle. Inspired also by Ludwig Wittgenstein's and Bertrand Russell's philosophical analyses of language, their 1929 manifesto, *Wissenschaftliche Weltauffassung* (Scientific World-Conception), named Albert Einstein as the key scientific inspiration for a positivistic worldview that banished metaphysics from science in favor of sense impressions arranged according to the principles of modern logic.

On November 28, 1930, Einstein responded to a letter from Schlick which tried to reconcile Einstein's present unease regarding the dominant interpretation of quantum mechanics with Einstein's earlier admiration of Mach. Einstein forcefully disagreed with Schlick's understanding of his views:

> Generally considered your representation does not correspond to my form of opinion, since I find your entire notion too positivist, so to speak. Physics to be sure *supplies* relations between sense experiences, but only in a mediated manner. *Your position* for me in this respect turns out to be in no way exhaustive. I say this to you point-blank: Physics is an attempt at a conceptual construction of a model of *the real world* as well as its lawlike structure. Indeed it must exactly correspond to the empirical relations between the sense experiences accessible to us; but only *thus* is it linked to the latter.

Einstein insisted that there was a genuine reality out there, not just representations of it; it was a reality that we comprehend by confronting it with theories that we have freely created. As he wrote in the conclusion to this letter: "You will be surprised by the 'metaphysician' Einstein. But every four- and two-legged animal is de facto a metaphysician in this sense."[5]

As he was entering his sixth decade, Einstein increasingly expressed a strongly realist understanding of nature not only in the philosophy of science but also when discussing religion and ethics. For example, in July 1930 he hosted the Bengali writer Rabindranath Tagore, laureate of the 1913 Nobel Prize in Literature, at his summer home at Caputh. Photographs of the meeting of these two luminaries were spread across the press, and Einstein's archives also contain an account of their conversation. Initially, Einstein posed brief questions to Tagore about his views of reality and beauty, which Tagore addressed in full paragraphs. Einstein became more animated when Tagore expressed the view that humanity can know nothing of either beauty or reality except from a subjective position because we lack access to both (a view related to Kant's). Einstein agreed with Tagore on the question of beauty but not about reality:

> I cannot *prove* that our scientific truth must be understood as a truth with content valid outside of human views, but I firmly believe it. I believe, for example, that the Pythagorean theorem of Geometry expresses something that is (approximately) true in a certain sense independent of the existence of people. In any event one can comprehend it as such. If there is a *reality* independent from people, so there is also a *truth* that corresponds to it; indeed the denial of the first engenders the denial of the existence of the second.[6]

The dispute was shortly disbanded with Einstein's comment: "I cannot prove that my view is correct, but it is my religion."

Einstein chose his words carefully. He did not mean "religion" in the sense of organized religion or communal identity. Einstein distinguished his comments on religion in general from statements about Judaism or Zionism, or other world religions. Indeed, he had harsh words for the world's clergy, who he felt "in the course of history have brought infinitely great struggle and war to humanity and thus bear great guilt."[7] Instead, as he told the *New York Times* in

April 1929, referring to the excommunicated seventeenth-century Dutch Jewish philosopher: "I believe in [Baruch] Spinoza's God, who reveals Himself in the orderly harmony of what exists, not in a God who concerns Himself with fates and actions of human beings."[8] Einstein used the word "God" to characterize the order found in nature and the sense of awe its elegance inspired in him. As he wrote in several articles for the American press, which frequently asked him about this topic, there could be no conflict between science and his version of religion since both were ways of appreciating that very thing he had described to Tagore as "reality." "Free creations of the human mind" were a religious notion for Einstein, but certainly not in the Judeo-Christian sense of "Creation."

After Einstein suddenly became famous in 1919 and, surprisingly, remained a global celebrity until his death in 1955, he was repeatedly asked to opine on such general questions. These philosophical pronouncements were reflections that had occupied him deeply for decades. The term "free creation," although not his only expression of his philosophy of science, cropped up in Einstein's thinking throughout his entire career: from an early letter to his first girlfriend, Marie Winteler, in 1896 down to his last interview days before his death with the historian of science I. B. Cohen. Einstein directed that final conversation to the work of Isaac Newton—the protagonist of his Spencer Lecture in 1933—since Cohen was a specialist on the early modern natural philosopher. Cohen was astonished by how Einstein thought of Newton almost like a peer: "Einstein apparently had little feeling for the way in which a man's mind is imprisoned by his culture and the character of his thoughts are molded by his intellectual environment. I did not press the point, but I was struck by the fact that in physics Einstein could see Newton as a man of the 17th century, but that in the other realms of thought and action he viewed each man as a timeless, freely acting individual to be judged as if he were a contemporary of ours."[9]

In his final years Einstein considered his own ideas, especially the triumph of relativity theory and his continued pondering over quantum mechanics, as free creations of the human mind and therefore in some sense timeless. As support for his position, one might observe that his ideas continue to animate our own moment as well as they did his own a century ago. Taking a somewhat arbitrary metric of the Nobel Prizes in Physics awarded in the twenty-first century, no fewer than seven (2001, 2006, 2011, 2017, 2019, 2020, and 2022) stem directly from Einstein's work in 1905 and 1915. That said, he was, as Cohen observed, a man of his time, and we understand him better when we view him within his own patch of spacetime, while appreciating that view from our own.

NOTES

AEA   The **Albert Einstein Archives** at the Hebrew University in Jerusalem is the major repository of Einstein's papers, correspondence, and books. Each item in the collection has a unique identifier consisting of four to six numbers and is indicated in the notes with its location symbol AEA, followed by a number contained in square brackets.

CPAE   *The Collected Papers of Albert Einstein* is the authoritative published edition of Einstein's writings and correspondence. Each volume is available in a hardcover which includes the original texts, scholarly annotations, and references, and a softcover companion with translations into English. Seventeen volumes covering Einstein's life up to the year 1930 have so far been published by Princeton University Press and are available for free online at https://einsteinpapers.press.princeton.edu. In the notes, references are presented as X: Y, where X represents the volume number and Y the document number.

Prologue

1. "Prof. Einstein Here, Explains Relativity," *New York Times* (April 3, 1921): 1, 13, on p. 1.

2. Einstein, "Die Einwanderung aus dem Osten," reproduced in CPAE 7: 29.

3. Einstein to Fritz Haber, March 9, 1921, CPAE 12: 88.

4. *Berliner Illustrierte Zeitung*, no. 50 (December 14, 1919): cover.

5. Konrad Haenisch to Einstein, September 6, 1920, CPAE 10: 135.

6. Einstein to Paul Ehrenfest, September 9, 1920, CPAE 10: 139.

7. At a time when a typical Princeton professorial salary was $4,000 per year. Einstein to Fritz Haber, March 9, 1921, CPAE 14: 88.

8. "Raffiniert ist der Herrgott, aber boshaft ist er nicht." See CPAE 12, Calendar, entry of May 9, 1921, and Oswald Veblen to Einstein, April 17, 1930, CPAE 17: 324.

Chapter 1

1. Einstein to Conrad Habicht, May 18 or 25, 1905, CPAE 5: 27. Emphasis in original.

2. Max Planck to Robert W. Wood, October 7, 1931, in Archive for the History of Quantum Physics, Microfilm 66, 5, as cited in Thomas S. Kuhn, *Black-Body Theory and the Quantum Discontinuity, 1894–1912* (Chicago: University of Chicago Press, 1978), p. 132.

3. Max Planck, Nobel Lecture 1918 (delivered June 2, 1920), available at https://www.nobelprize.org/prizes/physics/1918/planck/lecture.

4. *Vorlesungen über die Theorie der Wärmestrahlung*, 1st ed. (Leipzig: Barth, 1906), p. 108n, as translated in Kuhn, *Black-Body Radiation*, p. 130.

5. Einstein, "Das Komptonsche Experiment: Ist die Wissenschaft um ihrer selbst willen da?" *Berliner Tageblatt* (April 20, 1924), reproduced in CPAE 14: 236.

6. Einstein to Habicht, undated but between June 30 and September 22, 1905, CPAE 5: 28.

Chapter 2

1. Max Planck, "Die Kaufmannschen Messungen der Ablenkbarkeit der β-Strahlen in ihrer Bedeutung für Dynamik der Elektronen," *Physikalische Zeitschrift* 7 (1906): 753–761, p. 756.

2. Felix Klein, "Über die geometrischen Grundlagen der Lorentzgruppe," *Deutsche Mathematiker-Vereinigung. Jahresbericht* 19 (1910): 281–300, p. 281.

3. Einstein to Michele Besso, March 1, 1916, CPAE 8: 178.

4. Galileo Galilei, *Dialogue Concerning the Two Chief World Systems*, tr. Stillman Drake (Berkeley: University of California Press, 1953), pp. 186–187.

5. Einstein foreword to Galileo, *Dialogue*, p. xvii.

6. All quotations from Einstein, "Vom Relativitäts-Prinzip," *Vossische Zeitung* (April 26, 1914), reproduced in CPAE 6: 1.

7. CPAE 6: 42, equation on p. 25 of the original, and p. 449 of CPAE.

8. Albert Einstein, "Zur Elektrodynamik bewegter Körper," *Annalen der Physik* 17 (1905): 891–921, reproduced in CPAE 2: 23, last line.

9. Albert Einstein, *Aether und Relativitätstheorie* (Berlin: Julius Springer, 1920), p. 9 and p. 12, reproduced in CPAE 7: 38.

10. Einstein to Edwin E. Slosson, undated but between June 26 and July 31, 1925, CPAE 15: 13.

11. Albert Einstein, "Meine Theorie und Millers Versuche," *Vossische Zeitung* (January 19, 1926), reproduced in CPAE 15: 161.

12. Albert Einstein, "Ist die Trägheit eines Körpers von seinem Energieinhalt abhängig?," *Annalen der Physik* 18 (1905): 639–641, on p. 641, reproduced in CPAE 2: 24.

13. Albert Einstein, "Über die vom Relativitätsprinzip geforderte Trägheit der Energie," *Annalen der Physik* 23 (1907): 371–384, on p. 382, n. 1, reproduced in CPAE 2: 45.

14. Quoted in Ronald W. Clark, *Einstein: The Life and Times* (New York: World Publishing Co., 1971), p. 537.

15. Albert Einstein, "Elementary Derivation of the Equivalence of Mass and Energy," *Bulletin of the American Mathematical Society* (April 1935): 223–230, p. 225.

16. Albert Einstein, "E = MC²: The Most Urgent Problem of Our Time," *Science Illustrated* 1, no. 1 (April 1946): 16–17, on p. 16.

17. Einstein, "E = MC²: The Most Urgent Problem of Our Time," p. 17.

18. Einstein, "Über das Relativitätsprinzip und die aus demselben gezogenen Folgerungen," *Jahrbuch der Radioaktivität und Elektronik* 4 (1907): 411–462, p. 442, reproduced in CPAE 2: 47.

19. Albert Einstein, "Grundgedanken und Methoden der Relativitätstheorie, in ihrer Entwicklung dargestellt," [after January 22, 1920], reproduced in CPAE 7: 31, p. 136 in English translation. Emphasis in original.

20. Albert Einstein, *Relativity: The Special and the General Theory*, tr. Robert W. Lawson (New York: Crown, 1961 [1917]), pp. 66–67.

21. "Einstein Expounds His New Theory," *New York Times* (December 3, 1919), p. 19.

22. Albert Einstein, "Time, Space, and Gravitation," *The Times (London)* (November 28, 1919), reproduced in CPAE 7: 26.

23. Jean Eisenstaedt, "The Low Water Mark of General Relativity, 1925–1955," in Don Howard and John Stachel, eds., *Einstein and the History of General Relativity* (Boston: Birkhäuser, 1989), pp. 277–292.

24. Einstein to P. Ehrenfest, undated but after October 22, 1929, CPAE 17: 96.

25. Einstein to G. Lemaître, September 24, 1947, AEA [15– 085.1].

Chapter 3

1. John F. Clauser, Nobel Prize Lecture, December 8, 2022, available at https://www.nobelprize.org/prizes/physics/2022/clauser/lecture.

2. Albert Einstein to Walter Dällenbach, July 1, 1919, CPAE 9: 66.

3. Albert Einstein to Max Born, June 4, 1919, CPAE 9: 56.

4. Paul Ehrenfest to Albert Einstein, September 16, 1925, CPAE 15: 68.

5. Albert Einstein to Paul Ehrenfest, September 18, 1925, CPAE 15: 71.

6. Albert Einstein to Hedwig and Max Born, December 30, 1921, CPAE 12: 345.

7. Quoted in CPAE 14: Introduction, p. xli.

8. CPAE 14: 225 and Introduction, p. lx.

9. CPAE 14: 240 and Introduction, p. lx.

10. Paul Ehrenfest to Albert Einstein, January 9, 1925, CPAE 14: 417.

11. Albert Einstein to Paul Ehrenfest, March 15, 1922, CPAE 13: 87, p. 107.

12. Einstein, "On the Crisis of Theoretical Physics," *Kaizo* 4, no. 12 (December 1922): 1–8, reproduced in CPAE 13: 318.

13. "Sound Recording for the Prussian State Library," February [6], 1924, CPAE 14: 208.

14. "Das Komptonsche Experiment: Ist die Wissenschaft um ihrer selbst willen da?" *Berliner Tageblatt* (April 20, 1924), CPAE 14: 236.

15. Albert Einstein to Max Born, December 4, 1926, CPAE 15: 426.

16. John L. Heilbron, "The earliest missionaries of the Copenhagen spirit," *Revue d'histoire des sciences* 38 (1985): 195–230, p. 197.

17. Niels Bohr, "The Atomic Theory and the Fundamental Principles Underlying the Description of Nature." Address to the 18th Meeting of

Scandinavian Scientists given on August 26, 1929. See Niels Bohr, *Collected Works*, ed. L. Rosenfeld (Amsterdam: Elsevier, 1985), Vol. 6, p. 253.

18. Niels Bohr, "Biology and Atomic Physics," *Celebrazione del secondo centenaio della nascita di Luigi Galvani, Bologna 18–21 ottobre 1937–XV: Rendiconto generale* (Bologna: Tipografia Luigi Parma, 1938), pp. 68–78. Also in Niels Bohr, *Collected Works*, 10: 49–62, on p. 60.

19. "Einstein Attacks Quantum Theory," *New York Times* (May 4, 1935): 11.

20. Albert Einstein and Nathan Rosen to Watson Davis, Director of Science Service, May 10, 1935, AEA [87-482].

21. A. Einstein, B. Podolsky, and N. Rosen, "Can Quantum Mechanical Description of Reality Be Considered Complete?" *Physical Review* 47 (May 15, 1935): 777–780, on p. 777. Emphasis in original.

22. Albert Einstein to Hendrik A. Lorentz, June 17, 1927, CPAE 16: 8.

23. November 3, 1927. See Jagdish Mehra and Helmut Rechenberg, *The Completion of Quantum Mechanics, 1926–1941* (New York: Springer, 2000), pp. 251–253, and CPAE 16: Introduction, p. xlvi.

24. Einstein exchange with Ehrenfest [October 25, 1927], CPAE 16: 75.

25. Erwin Schrödinger to Einstein, June 7, 1935, AEA [22-044].

26. Einstein to Schrödinger, June 19, 1935, AEA [22-047].

27. Wolfgang Pauli to Werner Heisenberg, June 15, 1935, in *Wissenschaftlicher Briefwechsel mit Bohr, Einstein, Heisenberg, a.o.*, ed. A. Hermann, K. v. Meyenn, V. F. Weisskopf (New York: Springer, 1979–2005), Vol. 2, pp. 402–409.

28. Pauli to Einstein, December 19, 1929, CPAE 17: 155.

29. Einstein to Pauli, December 24, 1929, CPAE 17: 171.

30. *Politiken*, October 7, 1935, pp. 89–90.

31. Niels Bohr, "Can Quantum Mechanical Description of Physical Reality Be Considered Complete?" *Physical Review* 48 (October 15, 1935): 696–702, on 702.

32. Albert Einstein to Paul Arthur Schilpp, February 5, 1948, AEA [80-482].

33. Neils Bohr, "Discussions with Einstein on Epistemological Problems in Atomic Physics," in Paul Arthur Schilpp, ed., *Albert Einstein: Philosopher–Scientist* (Evanston, IL, 1949), pp. 201–241, on p. 224, p. 228, pp. 236–237, and p. 239.

34. Einstein to Max Born, March 3, 1947, reproduced in Albert Einstein, Hedwig Born, and Max Born, *The Born-Einstein Letters 1916–1955* (New York: Macmillan, 2005), p. 154.

Chapter 4

1. Einstein to Willy Hellpach, October 8, 1929, CPAE 17: 79.

2. Einstein to Paul Nathan, April 3, 1920, CPAE 9: 366.

3. Einstein to Adolf Kneser, June 7, 1918, CPAE 8: 560.

4. *Deutsche Tageszeitung*, September 26, 1919, Morning Edition, p. [2].

5. To Chaim Weizmann, [Berlin, March 2, 1925], CPAE 14: 450.

6. To Chaim Weizmann, November 25, 1929, CPAE 17: 128.

7. Einstein to Maja Winteler-Einstein, March 2, 1930, CPAE 17: 274.

8. Einstein to Hans Albert Einstein [between August 1 and September 30, 1931]. AEA [75-714].

9. Einstein to Marcel Grossmann, September 12, 1920, CPAE 10: 148.

10. Einstein to Hans Albert Einstein [between August 1 and September 30, 1931]. AEA [75-714].

11. Einstein to Henri Barbusse, November 25, 1928, CPAE 16: 318.

12. *Jewish Daily Bulletin*, September 19, 1930, CPAE 17: 417.

13. Travel Diary for USA, December 3, 1931 to February 4, 1932, p. 4, entry of December 6, 1931. AEA [29-136].

14. First published March 11, 1933, in *New World Telegram*.

15. Einstein to Otto Hahn, January 28, 1948, quoted in Klaus Hentschel, *Die Mentalität deutscher Physiker in der frühen Nachkriegszeit (1945–1949)* (Heidelberg: Synchron, 2005), p. 159.

16. Einstein, "What I Believe," *The Forum* 84, no. 4 (October 1930): 103–104. CPAE 17: 425.

17. Quoted in Fred Jerome and Rodger Taylor, *Einstein on Race and Racism* (New Brunswick: Rutgers University Press, 2005), p. 137.

18. Quoted in Jerome and Taylor, *Einstein on Race and Racism*, pp. 140–141.

19. Einstein, "Why Socialism?," *Monthly Review* 1, no. 1 (1949): 9–15.

20. Fantova, "Gespräche mit Einstein," entries for April 13 and June 3, 1954. Hanna Fantova Collection, Princeton University Library C0703, Box 1, Folder 6.

21. Quoted in *TIME* (November 22, 1954).

22. Quoted in Yitzhak Navon, "On Einstein and the Presidency of Israel," in Gerald Holton and Yehuda Elkana, eds., *Albert Einstein: Historical and Cultural Perspectives* (Mineola, NY: Dover, 1982), pp. 293–296, on 294–295.

Chapter 5

1. Einstein to Helene Savić, after December 17, 1912, CPAE 5: 424.

2. "Military Service Book," CPAE 1: 91.

3. Einstein to Paul Ehrenfest, August 19, 1914, CPAE 8: 34.

4. Einstein, "In Support of Georg Nicolai," January 26, 1920, reproduced in CPAE 7: 32.

5. Einstein, "On the Questionnaire Concerning the Right of National Self Determination," undated but between July 1917 and March 10, 1918, CPAE 6: 45a.

6. Lecture Notes, Winter Semester 1918–1919, CPAE 7: 12.

7. Einstein to Kurt Hiller, August 21, 1931. AEA [46-693].

8. Raymond de Rienzi to Einstein, July 6, 1922, CPAE 13: 267.

9. Einstein to Heinrich Zangger, July 31, 1923, CPAE 14: 95.

10. Einstein, "Militant Pacifism—The Two Percent Speech," *World Tomorrow* 14 (1931): 9.

11. "Einstein on Arrival Braves Limelight for Only 15 Minutes," *New York Times* (December 12, 1930): 1, quotation on p. 16.

12. Einstein to Franklin D. Roosevelt, August 2, 1939, AEA [33-008].

13. "Einstein Bars Comment on Atomic Bomb at Present," *New York Times* (August 8, 1945): 9.

14. "Atoms Not Occult, Einstein Declares," *New York Times* (August 12, 1945): 29.

15. Henry DeWolf Smyth, *Atomic Energy for Military Purposes* (Princeton, NJ: Princeton University Press, 1945), p. 2.

16. Einstein to Raymond Swing, August 27, 1945, AEA [57-443].

17. Albert Einstein, "On the Atomic Bomb," *Atlantic Monthly* 176 (November 1945): 43–45.

18. Einstein, "On the Atomic Bomb."

19. Einstein to John Cromwell, May 22, 1946, and talk for Paramount newsreel, June 4, 1946. AEA [90-577] and [88-628].

20. "Einstein Deplores Use of Atom Bomb," *New York Times* (August 19, 1946): 1.

21. See Einstein typescript for interview, AEA [28-870] and *New York Times*, February 13, 1950.

22. Quoted in CPAE 13, Introduction, p. lvii.

23. Einstein, "On My Participation in the Atom Bomb Project," *Kaizo*, Special Edition on the Atomic Bomb, November 1952.

24. Einstein to Seiei Shinohara, February 28, 1953, AEA [61-295].

25. Seiei Shinohara to Einstein, June 18, 1953. AEA [61-296].

26. "Einstein-Russell Manifesto," July 9, 1955, available at https://pugwash. org/1955/07/09/statement-manifesto.

Chapter 6

1. Einstein, "Reply to Criticisms," in Paul Arthur Schilpp, ed., *Albert Einstein: Philosopher-Scientist* (La Salle, IL: Open Court, 1970 [1949], p. 684.

2. Einstein, "On the Method of Theoretical Physics," *Philosophy of Science* 1, no. 2 (April 1934): 163–169, on p. 165.

3. Einstein, "On the Method of Theoretical Physics," p. 166.

4. Einstein, "On the Method of Theoretical Physics," p. 167.

5. Einstein to Moritz Schlick, November 28, 1930, CPAE 17: 497. Einstein's emphasis.

6. Einstein, "On the Nature of Reality," July 14, 1930, CPAE 17: 372. Einstein's emphasis.

7. Einstein to Friedrich Siegmund-Schultze, May 23, 1930, CPAE 17: 339.

8. "Einstein Believes in 'Spinoza's God,'" *New York Times* (April 25, 1929): 6. Reproduced in CPAE 16: 508.

9. I. Bernard Cohen, "The Last Interview," *Scientific American* 193, no. 1 (July 1955): 68–73, p. 72.

FURTHER READING

EINSTEIN'S WRITINGS

Calaprice, Alice, ed. *The Ultimate Quotable Einstein*. Princeton, NJ: Princeton University Press, 2010.

Dukas, Helen, and Banesh Hoffmann, eds. *Albert Einstein, The Human Side: Glimpses from His Archives*. Princeton, NJ: Princeton University Press, 2014.

Einstein, Albert. *Ideas and Opinions*. New York: Crown Publishers, 1954 (and many later editions).

Einstein, Albert. *The Meaning of Relativity*. Princeton, NJ: Princeton University Press, 2014 [1945].

Einstein, Albert. *Out of My Later Years*. New York: Philosophical Library, 1950.

Einstein, Albert. *Relativity: The Special and the General Theory*. 100th anniversary edition, ed. Hanoch Gutfreund and Jürgen Renn. Princeton, NJ: Princeton University Press, 2015.

Renn, Jürgen, and Robert Schulmann, eds. *Albert Einstein, Mileva Marić: The Love Letters*. Princeton, NJ: Princeton University Press, 2000.

Rosenkranz, Ze'ev, ed. *The Travel Diaries of Albert Einstein: South America, 1925*. Princeton, NJ: Princeton University Press, 2023.

Rosenkranz, Ze'ev, ed. *The Travel Diaries of Albert Einstein: The Far East, Palestine, and Spain, 1922–1923*. Princeton, NJ: Princeton University Press, 2018.

Schilpp, Paul Arthur, ed. *Albert Einstein: Philosopher-Scientist: The Library of Living Philosophers Volume VII*. La Salle, IL: Open Court, 1970 [1949].

Stachel, John, ed. *Einstein's Miraculous Year: Five Papers That Changed the Face of Physics*. Princeton, NJ: Princeton University Press, 1998.

BIOGRAPHIES

Clark, Ronald W. *Einstein: The Life and Times.* New York: Harper Paperbacks, 2011 [1971].

Fölsing, Albrecht. *Albert Einstein: A Biography.* New York: Viking, 1997.

Frank, Philipp. *Einstein: His Life and Times.* New York: Alfred A. Knopf, 1953.

Isaacson, Walter. *Einstein: His Life and Universe.* New York: Simon & Schuster, 2008.

Moszkowski, Alexander. *Einstein, the Searcher: His Work Explained from Dialogues with Einstein.* New York: Methuen & Co., 1921.

Pais, Abraham. *Subtle Is the Lord: The Science and the Life of Albert Einstein.* New York: Oxford University Press, 1982.

BIOGRAPHICAL COLLECTIONS

Calaprice, Alice, Daniel Kennefick, and Robert Schulmann, eds. *An Einstein Encyclopedia.* New York: Princeton University Press, 2016.

Holton, Gerald, and Yehuda Elkana, eds. *Albert Einstein: Historical and Cultural Perspectives: The Centennial Symposium in Jerusalem.* Princeton, NJ: Princeton University Press, 1982.

Janssen, Michel, and Christoph Lehner, eds. *The Cambridge Companion to Einstein.* Cambridge: Cambridge University Press, 2014.

Stachel, John. *Einstein from 'B' to 'Z'.* Boston: Birkhäuser, 2002.

RELATIVITY

Buchwald, Jed Z., ed. *Einstein Was Right: The Science and History of Gravitational Waves.* Princeton, NJ: Princeton University Press, 2020.

Janssen, Michel, John D. Norton, Jürgen Renn, Tilman Sauer, and John Stachel, eds. *The Genesis of General Relativity: Sources and Interpretations,* 4 vols. Dordrecht: Springer, 2007.

Kennefick, Daniel. *No Shadow of a Doubt: The 1919 Eclipse That Confirmed Einstein's Theory of Relativity.* Princeton, NJ: Princeton University Press, 2021.

Kennefick, Daniel. *Traveling at the Speed of Thought: Einstein and the Quest for Gravitational Waves.* Princeton, NJ: Princeton University Press, 2007.

Kox, Anne J., and Henrietta F. Schatz. *A Living Work of Art: The Life and Science of Hendrik Antoon Lorentz.* New York: Oxford University Press, 2021.

Norton, John D. *Einstein for Everyone.* Nullarbor Press, 2021 at https://sites.pitt.edu/~jdnorton/teaching/HPS_0410/chapters)

Staley, Richard. *Einstein's Generation: The Origins of the Relativity Revolution.* Chicago: University of Chicago Press, 2008.

QUANTUM THEORY

Baggott, Jim, and John L. Heilbron. *Quantum Drama. From the Bohr-Einstein Debate to the Riddle of Entanglement.* Oxford: Oxford University Press, 2024.

Beller, Mara. *Quantum Dialogue: The Making of a Revolution.* Chicago: University of Chicago Press, 2001.

Cassidy, David B. *Uncertainty: The Life and Science of Werner Heisenberg.* New York: W. H. Freeman, 1993; and the updated abridgement *Beyond Uncertainty: Heisenberg, Quantum Physics, and the Bomb.* New York: Bellevue Press, 2009.

Fine, Arthur. *The Shaky Game: Einstein, Realism, and the Quantum Theory.* Chicago: University of Chicago Press, 1986.

Heilbron, John L. *The Dilemmas of an Upright Man: Max Planck as Spokesman for German Science.* Cambridge, MA: Harvard University Press, 1996.

Heilbron, John L. "The Earliest Missionaries of the Copenhagen Spirit." *Revue d'histoire des sciences* 38 (1985): 195–230.

Heilbron, John L. *Niels Bohr: A Very Short Introduction.* New York: Oxford University Press, 2020.

Klein, Martin J. "Max Planck and the Beginnings of Quantum Theory." *Archive for History of the Exact Sciences* 1, no. 5 (1961): 459–479.

Klein, Martin J. *Paul Ehrenfest: The Making of a Theoretical Physicist.* London and New York: North-Holland Publishing Company, 1970.

Kuhn, Thomas S. *Black-Body Theory and the Quantum Discontinuity, 1894–1912.* Chicago: University of Chicago Press, 1987 [1978].

Norton, John D. "Atoms, Entropy, Quanta: Einstein's Miraculous Argument of 1905." *Studies in History and Philosophy of Modern Physics* 37 (2006): 71–100.

Stone, A. Douglas. *Einstein and the Quantum: The Quest of the Valiant Swabian.* Princeton, NJ: Princeton University Press, 2013.

EINSTEIN'S CULTURAL AND POLITICAL ENGAGEMENTS

Boyer, Paul S. *By the Bomb's Early Light: American Thought and Culture at the Dawn of the Atomic Age.* Chapel Hill: University of North Carolina Press, 1994 [1985].

Gordin, Michael D. *Einstein in Bohemia*. Princeton, NJ: Princeton University Press, 2022.

Grundmann, Siegfried. *The Einstein Dossiers: Science and Politics—Einstein's Berlin Period with an Appendix on Einstein's FBI File*, tr. Ann M. Hentschel. Berlin: Springer, 2005.

Hentschel, Klaus, ed., and Ann M. Hentschel, tr. *Physics and National Socialism: An Anthology of Primary Sources*. Basel: Birkhäuser Verlag, 1996.

Jammer, Max. *Einstein and Religion*. Princeton: Princeton University Press, 2002.

Jerome, Fred. *The Einstein File: J. Edgar Hoover's Secret War Against the World's Most Famous Scientist*. New York: St. Martin's Press, 2002.

Jerome, Fred, and Rodger Taylor. *Einstein on Race and Racism*. New Brunswick, NJ: Rutgers University Press, 2005.

Nathan, Otto, and Heinz Norden, eds. *Einstein on Peace*. New York: Simon and Schuster, 1960.

Robinson, Andrew. *Einstein on the Run: How Britain Saved the World's Greatest Scientist*. New Haven, CT: Yale University Press, 2019.

Rosenkranz, Ze'ev. *Einstein Before Israel: Zionist Icon or Iconoclast?* Princeton, NJ: Princeton University Press, 2011.

Schulmann, Robert, and David E. Rowe, eds. *Einstein on Politics: His Private Thoughts and Public Stands on Nationalism, Zionism, War, Peace, and the Bomb*. Princeton, NJ: Princeton University Press, 2007.

Smith, Alice Kimball. *A Peril and a Hope: The Scientists' Movement in America, 1945–1947*. Chicago: University of Chicago Press, 1965.

Stanley, Matthew. *Einstein's War: How Relativity Triumphed Amid the Vicious Nationalism of World War I*. New York: Dutton, 2019.

Stern, Fritz. *Einstein's German World*. Princeton, NJ: Princeton University Press, 1999.

INDEX

For the benefit of digital users, indexed terms that span two pages (e.g., 52–53) may, on occasion, appear on only one of those pages.

Aargau Kantonsschule, 75
absolute space, 109–10
African Americans, 84–86
American Crusade to End
    Lynching, 85
American Mathematical Society, 39
Anderson, Marian, 85
Annalen der Physik, 16, 20–21,
    31–32, 38
Anschutz-Kaempfe, 80
antisemitism
    attacks on Einstein and relativity
        theory, 2, 3, 4–6, 37–38, 41
    Einstein's exposure of and stand
        against, 5–6, 11
    Einstein's identity and, 70
    Einstein's refugee
        collaborators, 26–27
    Einstein's schooling in
        Munich, 73–74
    Holocaust, 82–83

Mussolini's Italy, 24
    as a problem of non-Jews, 70, 85
    Russia and Soviet Union, 86
Aspect, Alain, 68
Atlantic Monthly, 100–101
Atombau und Spektrallinien (*Atomic
    Structure and Spectral Lines*)
    (Sommerfeld), 51–53
Atomic Energy Commission, 87–88
Atomic Energy for Military Purposes
    (Smyth), 99–100
atomic scientists' movement, 101–102
atomism, 20, 48, 110
Aydelotte, Frank, 23

Baha'i movement, 95–96
Baruch, Bernard, 102–3
Belgenland (ship), 96
Bell, John, 67–68
Ben-Gurion, David, 88
Bergmann, Hugo, 3, 79

Bergmann, Peter, 26–27
Berks, Robert, 83–84
Berlin Academy, *See* Prussian
    Academy of Sciences
Berliner Tageblatt, 2, 73–74
Besso, Michele, 11–13, 28, 30, 36
Bethe, Hans, 101–2
Big Bang, 46
binary star motion, 26
black bodies, 16–17
Blumenfeld, Kurt, 3
Bohr, Niels, 25–26, 50, 51–53, 52*f*,
    54–55, 56–58, 59–60, 61–62,
    63–68, 81
Bohr–Kramers–Slater paper, 54–55
Bolshevik Revolution, 76–77, 86
Boltzmann, Ludwig, 17
Born, Hedwig, 54–55
Born, Max, 51–53, 54–55, 56,
    60–61, 92
Bothe, Walther, 53, 54–55
Briggs, Lyman J., 98
British Royal Astronomical
    Society, 4
Brit Shalom, 79
Broglie, Louis de, 59
Brooklyn Jewish Hospital, 64–65
Buek, Otto, 91–92
Bund Neues Vaterland, 6, 92, 94
Bureau of Standards, 98
Bush, Vannevar, 98–99

Cairncross, John, 98–99
California Institute of Technology
    (Caltech), 22–23, 25, 46–47,
    81–82, 84, 96, 106
"Can Quantum-Mechanical
    Description of Physical Reality
    Be Considered Complete?"
    (Einstein–Podolsky–Rosen

[EPR] paper), 25–26, 58–59,
    60–62, 63, 67–68
Case School of Applied Science,
    6–8, 37–38
Cassirer, Ernst, 4–5
Central Organization for a Durable
    Peace, 92
Chamberlain, Austen, 107*f*
Charles-Ferdinand University,
    9–10, 26
civil rights movement, 84–86
Clauser, John, 68
Clausius, Rudolf, 16–17
Cohen, I. B., 113–14
Columbia University, 22–23
Communism, 26, 85–88, 96
complementarity principle, 49–50,
    56–57, 58, 66
Compton, Arthur H., 54, 55–
    56, 59–60
Compton effect, 54–56
Conant, James Bryant, 98–99
Condon, Edward, 39, 101–2
Copenhagen interpretation, 25–26,
    56–57, 62, 66
correspondence principle, 66
cosmological constant, 46–47
Crisis, The, 84
Cultural History (Friedell), 81
Curie, Marie, 95*f*

dark energy, 46–47
Debye, Peter, 51–53
Dedekind, Richard, 12–13
Degenhart, Joseph, 75
*Dialogue Concerning the Two Chief
    World Systems* (Galileo),
    4–5, 31–32
Diecke, Gerhard, 59–60
Dirac, Paul, 23

"Discussion with Einstein on Epistemological Problems in Atomic Physics" (Bohr), 65
"Does Science Exist for Its Own Sake?" (Einstein), 55–56
"Does the Inertia of a Body Depend upon Its Energy Content?" (Einstein), 21, 38
Du Bois, W. E. B., 84
Dukas, Helen, 24, 76f

Eban, Abba, 88
Eddington, Arthur Stanley, 3–4, 44, 45f
Ehrenfest, Paul, 5–6, 45f, 46–47, 51–53, 59–60, 90–91
Eidgenössische Technische Hochschule (ETH; Federal Institute of Technology), 9–10, 11–14, 26–27, 51, 75
Einstein, Albert. See also quantum theory; relativity theory
  aiding refugees, 24–25, 76–77, 82, 107f
  aneurysm, 64–65
  antisemitism, 2, 3, 4–6, 11, 37–38, 41, 70, 73–74, 85
  atomism, 20, 48
  Bavaria, 71–74
  Berlin, 9–10, 22–23, 25, 58, 70, 75–76, 76f, 89–91, 92, 93–94
  Bern, 10, 11, 13–14, 21–22, 28
  birth of, 9, 71, 73
  birth of children, 13
  boycott of German scientists and reestablishment of international scientific relations, 4, 6, 82, 91–92
  Bund Neues Vaterland, 6, 92, 94

Caltech, 22–23, 81–82
childhood of, 71, 73
citizenship, 11–12, 24, 69–70, 89
civil rights movement, 84–86
Communism and McCarthyism, 86–88
criticism of and attacks on, 2, 4–5, 6–8, 27, 37–38, 53, 61
as cultural icon, 69–70
death of, 24–25
death threats, 94, 106
decision to stay in America, 23–24, 82
English language, 26–27, 39, 60–61, 83, 106
epistemology and the philosophy of science, 12–13, 58, 65, 66, 67, 107–14
ETH, 9–10, 11–14, 26–27, 51, 75
family electrical manufacturing business, 11–12, 72
FBI investigation, 86, 96
first marriage of, 13–14, 89–90
free creations of the human mind, 108–9, 112–14
geographically-related personas of, 9–11
Hebrew University, 1–2, 3, 77, 78f
honorary degrees, 21–22, 85
Institute for Advanced Study, 22–24, 67
internationalism, 4, 6, 58, 82, 92–93, 94
League of Nations, 92–93, 95f, 95
letter writing, 70
logical empiricism, 110
mad defenders of principles versus virtuosos, 51–53

Einstein, Albert (*cont.*)
  Milan, 11–12, 74–75
  Nazism, 82–83
  neo-Kantians, 110
  Nobel Prize, 19–20, 21–22, 50
  nuclear weapons, 40–41, 86, 90, 97–101, 102–5
  offer to become president of Israel, 88
  Olympia Academy, 12–13, 28, 110
  Oxford University, 22–23, 58, 106–8, 109–10
  pacifism, 6, 11, 24, 58, 79–80, 89–96, 101, 103–4
  papers written during *annus mirabilis*, 14–16, 20–21, 49
  Patent Office, 10, 12, 21–22
  photographs of, 7*f*, 45*f*, 52*f*, 72*f*, 76*f*, 78*f*, 94*f*, 107*f*
  Prague, 10, 43–44
  pressure to leave Germany, 2, 3–6, 22, 80, 81, 93
  Princeton University, 6–8, 10–11, 22–27, 28, 82, 83, 85, 106
  publishing corrections, 53
  realism, 111–12
  relationship with democracy and American politics, 83–88
  relationship with Jewish community, 1–3, 24–25, 70, 75–79, 82, 85
  religion, 70, 112–13
  resistance to Germany's claim on, 75–76, 80
  schooling, 9, 73–75
  sculpture of, 83–84
  second marriage, 24
  socialism, 86
  sound recording, 55
  speed of light, 6–8, 32–33
  Swiss draft board, 89
  trajectory of Mercury, 3–4, 44
  ulcerative colitis and exhaustion, 75–76
  unified field theory, 27, 28, 48, 58, 61, 106–7
  wartime manifesto and counter-manifesto, 91–92
  World War I, 89–92
  "young Einstein" versus "old Einstein," 10–11, 28
  Zionism, 1–3, 6, 11, 76–80
  Zurich, 9–10, 11–12, 13, 14*f*, 21–22, 75–76, 89–90
Einstein, Eduard, 13, 14*f*, 22–23, 89–90
Einstein, Elsa, 2, 24, 75–76
Einstein, Hans Albert, 13, 14*f*, 89–90
Einstein, Hermann, 11–12, 71–72, 74
Einstein, Jakob, 12, 72, 74
Einstein, Lieserl, 13
Einstein & Garrone Company, 74
Einstein-Knecht, Frieda, 14*f*
Einstein Marić, Mileva, 11–12, 13–14, 14*f*, 22–23, 24, 75–76, 89–90
Einstein–Podolsky–Rosen (EPR) paper ("Can Quantum-Mechanical Description of Physical Reality Be Considered Complete?"), 25–26, 58–59, 60–62, 63, 67–68
"Einstein-Russell Manifesto," 104–5
Einstein Tower, 44
"Electrodynamics of Moving Bodies" (Einstein), 31–32, 34–35

"Elementary Derivation of the Theorem Concerning the Equivalence of Mass and Energy, An" (Willard Gibbs Lecture) (Einstein), 39
Emergency Committee of Atomic Scientists, 101–3
entanglement paradox, 59, 67–68
entropy, 16–18
epistemology and the philosophy of science, 12–13, 57, 58, 65, 66, 67, 107–14
EPR (Einstein–Podolsky–Rosen) paper ("Can Quantum-Mechanical Description of Physical Reality Be Considered Complete?"), 25–26, 58–59, 60–62, 63, 67–68
equivalence of mass and energy $(E = mc^2)$, 21, 38–41, 99–100
equivalence principle, 42–43
ETH (Eidgenössische Technische Hochschule; Federal Institute of Technology), 9–10, 11–12, 13–14, 26–27, 51, 75
"Ether and the Theory of Relativity" (Einstein), 37
Evolution of Physics, The (Einstein and Infeld), 26

Falastin, 79
falling-man thought experiment, 42–43
Fantova, Hanna, 87
Federal Bureau of Investigation (FBI), 86, 96
Federal Institute of Technology (Eidgenössische Technische Hochschule [ETH]), 9–10, 11–12, 13–14, 26–27, 51, 75

Fermi, Enrico, 97–98
field theory, 26, 53, 57
Fifth Solvay Conference, 52*f*, 59, 65
Fine, Henry Burchard, 7*f*
Flexner, Abraham, 22–24
Förster, Wilhelm J., 91, 93
Franz Ferdinand, 89–90
Frauenglass, William, 86–87
free creations of the human mind, 108–9, 112–14
Freundlich, Erwin F., 4
Friedell, Egon, 81
Friedmann, Alexander, 46
Frisch, Otto, 98–99

Galileo Galilei, 4–5, 31–32
Gehrcke, Ernst, 4–5
Geiger, Hans W., 53, 54–55
general relativity. *See* relativity theory
Geodetic Institute, 93
"Geometry and Experience" (Einstein), 108–9
Georgia Tech, 83–84
Gerlach, Hellmut von, 93
German Physical Society, 5–6
German Society for Ethical Culture, 91
Gödel, Kurt, 23–24
Goudsmit, Samuel, 59–60
gravitational lensing, 25–26, 44, 47
gravitational redshift, 44
gravitational waves, 25–26, 29, 35, 47
"Greeting to America" (Einstein), 96
Gromyko, Andrei, 102–3
Grossmann, Marcel, 12, 26–27, 43–44
Grünberg, Theodor Koch, 81

Gumbel, Emil J., 93

Haber, Fritz, 6, 54–55, 90–91
Habicht, Conrad, 12, 15, 21
Hahn, Otto, 82–83
Harding, Warren G., 6
Harvard University, 44, 99
Hebrew University, 1–2, 3, 77, 78*f*
Heine, Wolfgang, 76–77
Heisenberg, Werner, 56–57, 61–62
Heisenberg's uncertainty
    principle, 59, 66
Helmholtz, Hermann von, 12–13
Helpach, Willy, 70
Herbert Spencer Lecture ("On
    the Method of Theoretical
    Physics") (Einstein), 58,
    106–7, 109–10
Hersey, John, 102
Hertz, Heinrich, 18
Hiller, Kurt, 92–93
Hiroshima (Hersey), 102
Hitler, Adolf, 41, 69–70, 80, 81–82,
    89–90, 106, 108–9
Hoffmann, Banesh, 26–27
Holocaust, 82–83
Hoover, J. Edgar, 86
House Un-American Activities
    Committee, 86–87
Hoyle, Fred, 46
Hubble, Edwin, 46–47
Hume, David, 12–13
Hylan, John F., 2

"Immigration from the East"
    (Einstein), 76–77
Infeld, Leopold, 26
Institute for Advanced Study (IAS),
    22–24, 25, 67, 87
interferometers, 35–36, 44

International Commission on
    Intellectual Cooperation, 95*f*, 95
International Congress of
    Physicists, 56
internationalism, 4, 6, 58, 82,
    92–93, 94
International Union of Pure and
    Applied Physics, 14–15
Israel Academy of Sciences and
    Humanities, 83–84
Israel and Levi featherbedding
    business, 71

*Jahrbuch der Radioaktivität und
    Elektronik*, 41
Japan Academy, 103
J. Einstein & Cie. electrical
    factory, 72
Jewish community, 1–2
    Einstein's family origins, 71
    Einstein's relationship to, 1–3,
        24–25, 70, 75–79
    Hebrew University, 1–2, 3,
        77, 78*f*
    Holocaust, 82–83
    offer to Einstein to become
        president of Israel, 88
    persecution and refugee crisis, 2,
        24–25, 76–77, 82, 88
    violent demonstrations in
        Palestine (1929), 77
    Zionism, 1–3, 6, 11, 76–80
Jewish Haganah, 77
Joliot-Curie, Frédéric, 97–98
Jordan, Pascual, 56
Journal of the Franklin Institute, 47

Kafka, Franz, 10
Kaiser Wilhelm Institute for
    Theoretical Physics, 9–10

Kaizo journal, 103–4
Kant, Immanuel, 110
Klein, Felix, 29–30
Koch, Pauline, 71
Kramers, Hendrik, 54–55
Kristallnacht, 82–83, 88

Langevin, Paul, 94*f*, 94
Laser Interferometric
    Gravitational-Wave
    Observatory (LIGO), 35
Laue, Max von, 4–5, 61
Laurence, William L., 62
League of Nations, 4, 92–93, 95*f*, 95
*Lectures on the Theory of Heat
    Radiation* (Planck), 17–18
Lemaître, Georges, 46–47
Lenard, Philipp, 4–5, 18
Leiden Observatory, 45*f*
Library of Living Philosophers
    (Schilpp), 63–64
Liebknecht, Karl, 93
light. *See also* quantum theory
    equivalence of mass and
        energy, 40
    relativity and, 32–34
    speed of, 6–8, 32–33, 34, 39, 40
    wave theory, 19, 53, 54
LIGO (Laser Interferometric
    Gravitational-Wave
    Observatory), 35
Lincoln University, 85
Locker-Lampson, Oliver S., 107*f*
logical empiricism, 110
Lorentz, Hendrik A., 4–5, 31,
    36–37, 45*f*, 51–53
Löwenthal, Ilse, 24, 75–76
Löwenthal, Margot, 24, 75–76
Löwenthal, Max, 24
Luitpold Gymnasium, 73–74, 75

luminiferous ether, 6–8, 20–
    21, 35–38
Luxemburg, Rosa, 93

Mach, Ernst, 12–13, 36, 110
magnet-coil thought
    experiment, 34, 36
Manhattan Project, 62, 98–100,
    101, 104–5
"Manifesto to the Europeans"
    (counter-manifesto),
    91–92, 93
Maxwell's equations, 19–20,
    21, 33–34
Mayer, Walther, 25, 26–27
McCarter Theatre, 85
McCarthyism, 86–88
Meaning of Relativity, The
    (Einstein's Princeton
    lectures), 6–8
Mercury, trajectory of, 3–4, 44
Michelson, Albert W., 35–36, 37–38
"Militant Pacifism" ("Two-Percent"
    speech) (Einstein), 95–96, 99
Miller, Dayton C., 6–8, 37–38
Millikan, Robert A., 95*f*
Monthly Review, The, 86
Morley, Edward W., 35–36, 37–38
Mount Wilson Observatory,
    6–8, 44
Munich International Electrical
    Exhibition, 72
muons, 33–34
Mussolini, Benito, 24

Nashashibi, Azmi al-, 79
Nathan, Paul, 73–74
National Association for the
    Advancement of Colored
    People (NAACP), 84

National Broadcasting Corporation (NBC), 96, 103
National Research Defense Committee, 99
Nature, 61–62
Nazism, 11, 18, 22–23, 41, 69–70, 76–77, 80, 81–83, 89–90, 97–98, 106, 108–9
"Negro Question, The" (Einstein), 84–85
neo-Kantians, 110
Nernst, Walther, 4–5, 91
Neumann, John von, 23
"New Determination of Molecular Dimensions, A" (Einstein), 20
New History Society, 95–96
Newton, Isaac, 29, 42, 44, 109, 113
New Yorker magazine, 102
New York Times, 2, 3–4, 27, 44, 58, 62, 99, 102, 112–13
Nicolai, Georg F., 91–92, 93
Nissen, Rudolph, 64–65
Nobel Prizes, 19–20, 29, 35, 49, 50, 68, 112, 113–14
No More War! Movement, 94f, 94
nuclear fission and weapons, 40–41, 67, 86, 87, 90, 97–101, 103–5
Nuremberg Racial Laws, 82

Olympia Academy, 12–13, 28, 110
"On a Heuristic Point of View Concerning the Production and Transformation of Light" (Einstein), 16
"On the Einstein-Podolsky-Rosen Paradox" (Bell), 67–68
"On the Electrodynamics of Moving Bodies" (Einstein), 20–21, 38
"On the Investigation of the State of the Ether in a Magnetic Field" (Einstein), 75

"On the Method of Theoretical Physics" (Herbert Spencer Lecture) (Einstein), 58, 106–7, 109–10
"On the Theory of Brownian Motion" (Einstein), 20
Oppenheimer, Frank, 87
Oppenheimer, J. Robert, 23, 87–88
Ostwald, Wilhelm, 21–22
Oxford University, 22–23, 58, 106–8, 109–10

Paasche, Hans, 93
pacifism, 4, 6, 11, 24, 58, 79–80, 89–96, 101, 103–4
Patent Office, 10, 12, 21–22
Pauli, Wolfgang, 23, 61–62
Pearson, Karl, 12–13
Peierls, Fritz, 98–99
Perrin, Jean, 20
Petersschule, 73
photoelectric effect, 18–21, 29, 50, 83–84. *See also* quantum theory
Physical Review, 25–26, 47, 58, 61–62
Physikalisch-Technische Reichsanstalt, 53
Planck, Max, 16–18, 29–30, 91
Planck's law, 17–18, 19
Podolsky, Boris, 25–26, 60–61
Poincaré, Henri, 12–13, 31
Politiken, 63
Portland (ship), 81
positivism, 111
Pound, Robert, 44
Princeton University, 6, 22–27, 28, 82, 83, 99–100, 106
  civil rights movement, 85
  Einstein's collaborators, 25–27
  lectures, 6–8

Physics and Mathematics
    building, 7f
"young Einstein" versus "old
    Einstein," 10–11
Principia Mathematica
    (Newton), 29
Prussian Academy of Sciences,
    9–10, 16, 53, 108–9
Prussian State Library, 55

quantum mechanics. *See*
    quantum theory
quantum of action, 16, 17–18
quantum theory, 18–20, 22, 25–26,
    27, 28, 29, 49–68, 75–76
    characterizing the microworld
        through discrete
        quantities, 49–50
    complementarity principle,
        49–50, 56–57, 58, 66
    Compton effect, 54–56
    Copenhagen interpretation,
        25–26, 56–57, 62, 66
    correspondence principle, 66
    dice-playing comment, 56
    Einstein–Bohr debate, 50, 51–53,
        54–55, 57–58, 59–60, 61–62,
        63–66, 68, 106–7
    entanglement paradox, 59, 67–68
    EPR paper, 25–26, 58–59,
        60–62, 63, 67–68
    experiments with Geiger, 53
    post-WWII renewal of interest
        in, 67–68
    quantum superpositions, 62
    Schrödinger's cat, 62
    single-slit thought
        experiment, 63
Queen Mary (ship), 24

Rathenau, Walther, 95

Rebka, Glen, 44
relativity theory, 1, 2, 3–5, 6, 20–21,
    28, 29–48, 75–76
    cosmological constant and dark
        energy, 46–47
    cosmological expansion, 46–47
    criticisms of, 2, 4–5, 6–8, 37–38
    equivalence of mass and energy,
        21, 38–41, 99–100
    equivalence principle, 42
    falling-man thought
        experiment, 42–43
    Galileo, 31–32
    general relativity, 3–4, 20–21,
        28, 29, 30, 35, 37–38, 41,
        43, 44, 46, 47, 53–54, 68,
        75–76, 109–10
    gravitational lensing, 44, 47
    gravitational redshift, 44
    gravitational waves, 29, 35, 47
    light, 32–34
    luminiferous ether, 6–8,
        20–21, 35–38
    magnet-coil thought
        experiment, 34, 36
    myth of incomprehensibility, 30
    origin of term, 29–30
    quantum mechanics and, 68
    sound recording, 55
    space-chest thought
        experiment, 42–43
    space-time, 37, 39, 43–44, 47
    special relativity, 20–21, 28, 29,
        30, 31, 33, 34, 37–44
    time dilation, 33–34
    train-carriage thought
        experiment, 32, 35
*Relativity: The Special and the
    General Theory* (Einstein), 30
"Reply to Critics" (Einstein), 64–65
Riemann, Bernhard, 12–13

Robeson, Paul, 85

Roosevelt, Eleanor, 23, 103

Roosevelt, Franklin D., 23, 76–77, 97–98, 100, 102, 103–4

Rosen, Nathan, 25–27, 39, 47, 58

Rosenberg, Ethel, 86–87

Rosenberg, Julius, 86–87

Rotblat, Joseph, 104–5

Rotterdam (ship), 1

Royal Albert Hall, 107*f*

Rubens, Heinrich, 4–5

Rupp, Emil, 56

Russell, Bertrand, 103–5, 111

Rutherford, Ernest, 107*f*

Sachs, Alexander, 98

Schilpp, Paul Arthur, 63–64

Schlick, Moritz, 110–11

Schrödinger, Erwin, 56, 59, 60–61, 62

Science Illustrated, 39, 40–41

Scientific World-Conception (*Wissenschaftliche Weltauffassung*) (Vienna Circle manifesto), 111

separability principle (*Trennungsprinzip*), 60–61, 63, 65, 67–68

Shinohara, Seiei, 104

single-slit thought experiment, 63

Sitter, Willem de, 45*f*

Slater, John, 54–55

Smetskaia, Nadezhda, 13

Smyth, Henry DeWolf, 99–100

socialism, 86, 93

Society of German Natural Scientists and Physicians, 5–6

Solovine, Maurice, 12

Sommerfeld, Arnold, 4–5, 51–53, 54

space-chest thought experiment, 42–43

special relativity. *See* relativity theory

Spinoza, Baruch, 12–13, 112–13

Stalin, Joseph, 86, 98–99

Stark, Johannes, 41

St. John, Charles, 44

Swing, Raymond, 100

Szilard, Leo, 97–98, 101–2

Tagore, Rabindranath, 112–13

Teller, Edward, 98

tensor calculus, 43–44

thermodynamics, 12–13, 16–18, 110

time dilation, 33–34

TIME Magazine, 97

Times of London, 3–4, 44

"Today with Mrs. Roosevelt" (TV show), 103

"To the Civilized World" (manifesto), 91

train-carriage thought experiment, 32, 35

Trennungsprinzip (separability principle), 60–61, 63, 65, 67 68

Truman, Harry S., 102, 103

two-body problem, 25–26

"Two-Percent" speech ("Militant Pacifism") (Einstein), 95–96, 99

Uhlenbeck, George, 59–60

unified field theory, 27, 28, 48, 58, 61, 106–7

United Nations, 86, 88, 101, 102–3

University of Berlin, 3–4, 9–10, 16

University of Bern, 21
University of California, Berkeley, 87
University of Chicago, 97
University of Vienna, 111
University of Wisconsin, 6
University of Zurich, 20, 21–22
U.S. Army Corps of Engineers, 99

Veblen, Oswald, 6–8, 26
Versailles Peace Treaty, 3, 6, 80
Vienna Circle, 111
Vossische Zeitung, 32–33, 35, 37–38

wave theory, 19, 53, 54
Weber, H. F., 75
Weizmann, Chaim, 1–2, 79, 88
Wertheimer, Max, 92
Westmoreland (ship), 22
Weyl, Hermann, 23
Weyland, Paul, 4–5
"What I Believe" (Einstein), 84
Wheeler, John, 42–43
"Why Socialism?" (Einstein), 86

Wigner, Eugene, 98
Wilhelm I, 71
Willard Gibbs Lecture ("Elementary Derivation of the Theorem Concerning the Equivalence of Mass and Energy, An") (Einstein), 39
Winteler, Marie, 113
Winteler-Einstein, Maja, 24, 79–80
Winterthur technical school, 12
Wissenschaftliche Weltauffassung (Scientific World-Conception) (Vienna Circle manifesto), 111
Wittgenstein, Ludwig, 111
Women's League of America, 96
Working Association of German Natural Scientists for the Preservation of Pure Science, 4–5

Yukawa, Hideki, 104–5

Zeilinger, Anton, 68
Zionism, 1–3, 6, 11, 76–80